Monographien aus dem Gesamtgebiete der Psychiatrie

18

Psychiatry Series

Herausgegeben von
H. Hippius, München · W. Janzarik, Heidelberg
C. Müller, Prilly-Lausanne

Transmethylations and the Central Nervous System

Edited by
V. M. Andreoli A. Agnoli C. Fazio

With 45 Figures

Springer-Verlag
Berlin Heidelberg GmbH 1978

V. M. ANDREOLI, Ospedali Neuropsichiatrici di Verona, Verona, Italia

A. AGNOLI, Clinica Malattie Nervose e Mentali dell'Università di L'Aquila, L'Aquila, Italia

C. FAZIO, 1ª Clinica Malattie Nervose e Mentali dell'Università di Roma, Roma, Italia

Original edition: Transmetilazioni e sistema nervoso centrale. A cura di V. M. Andreoli, A. Agnoli, C. Fazio. Edizioni Minerva Medica, Torino/Italia, 1976

ISBN 978-3-642-88518-1   ISBN 978-3-642-88516-7 (eBook)
DOI 10.1007/978-3-642-88516-7

2123/3130-543210

Contents

VI

Introduction

V. M. ANDREOLI, A. AGNOLI, and C. FAZIO

The transfer of a methyl group from a donor to an acceptor compound is a fundamental biochemical process long since known to biologists; the process is involved, for instance, in the metabolism of porphyrins, nucleic acids, and fatty acids.

Only recently, however, did transmethylation processes reveal their decisive role in CNS biochemistry — namely with the discovery that such processes are linked with the biogenic amines that have been conclusively identified as the chemical mediators of neuronal transmission and, more broadly, of behavior.

The first suggestion that transmethylation processes might be involved in the origination of certain mental diseases came from Harley-Mason (1952), who noticed that many of the hallucinogenic substances known at that time contained methyl radicals, and particularly that mescaline represented the product of O-methylation of dopamine in positions 3, 4, and 5. This hypothesis was put forward when the O-methylation of catecholamines by catechol-O-methyltransferases had not yet been described.

Harley-Mason further proposed that in the living organism the process of O-methylation might follow a "deviant" metabolic pathway, and that the accumulation of abnormal methylated metabolites endowed with hallucinogenic properties might be responsible for the implementation of some mental diseases, notably schizophrenia.

He also called attention to a substance, 3,4-dimethoxyphenylethylamine (3,4—DMPEA) as a possible psychotogenic molecule, on the strength of its capacity for producing catatonia in experimental animals.

In this respect we must mention that as far back as 1921 De Jong had done a number of studies on experimental catatonia, and had found that many of the numerous chemicals capable of causing that reaction were methyl or methoxyl compounds: thus epinephrine, acetylcholine, and mescaline. De Jong's results were published in book form in 1945, and they must have influenced later theories on the pathogenesis of schizophrenia: from that of Osmond and Smythies (1952) advocating the M substance, or mescalinelike substance, to that of Hoffer (1954) and that of Friedhoff and van Winkle (1962) —all of which involved methylated or methoxylated substances.

This "transmethylation theory" was invoked again in more recent times, to account for the manifestations of amphetamine-induced psychosis, in so many ways similar to those of acute paranoid schizophrenia.

The importance later attributed to catecholamines in the mediation of behavior, and the structural affinities between these physiologic CNS substances and 3,4-DMPEA, mescaline, and amphetamine itself, all of which are more or less markedly psychotogenic, again suggested that a metabolic error and the resulting accumulation of methoxylated products might play a major role in the regulation of behavior and in the origination of mental disorders.

Later still, similar transmethylation reactions were found to involve serotonin — another mediator of behavior. This indole compound was shown to produce psychotogenic methylated derivatives such as bufotenin and psilocybin; and as structural similarities were something of a craze at the time, resemblances were happily discovered even between serotonin and lysergic acid derivatives.

Now all these data are part and parcel of the recent history of biological psychiatry, and at the same time recall some more ancient but very brilliant anticipations, which had at least the merit of stimulating the development of behavioral neurochemistry. Today we can claim that we know a little more about the kinetics of transmethylation processes, in the sense that a number of enzymes that catalyze these processes have been identified; but we know much less about the etiologic, pathogenetic, and clinical meaning of these processes in actual neurologic and psychiatric disorders, as the following will exemplify. In 1969, Buscaino found that the blood serum of schizophrenic patients would yield a significantly greater than normal amount of methyl radicals to such substrates as nicotinic acid and nicotinamide (both being well-known acceptors of methyl groups). This was taken as the first experimental contribution in support of the transmethylation hypothesis: we had proof that hypermethylation was possible, and this gave added strength to the idea that such methyl and/or methoxy-derivatives as had been proposed in theory could really form through endogenous processes; in particular, schizophrenia could conceivably be associated with an exaggerated methylation of endogenous substrates such as the catecholamines or the indoleamines. But then in a recent paper (1975) Levi and Waxmann claimed the exact opposite, namely that schizophrenia is characterized by a blockade of methylation processes, since S-adenosyl-L-methionine (SAMe) is apparently synthesized at a lower turnover rate.

This methyl donor would in some way constitute a limiting factor to normal transmethylation processes — to the point where the authors recommend the administration of methyl donors for therapeutic purposes. In this present volume you will find data on the SAMe content of serum from acute and chronic schizophrenics, showing a reduction of about 50% in acute forms. This finding may be interpreted according to Buscaino (excess transmethylation, hence increased consumption of donor substance) or to Levi-Waxmann (blockade of donor synthesis, hence reduction of biological transmethylation processes). All this goes to show that it is still too soon to hope for a rational explanation of behavioral disorders based on experimental findings concerning the biochemical model of transmethylation: the whole problem is wide-open and constitutes for the time being nothing more than a sound working hypothesis.

This book is not a history of hypotheses concerning transmethylation processes, but a synopsis of research work on such processes in the central nervous system — or if you will, a summary of the role of these biochemical operations in the brain. Still, the book would probably not exist if we had not also looked at these reactions with a clinician's eye and thus commanded a view of the problems, involved including the clinical, the experimental, and the therapeutic angles.

The risk of working exclusively on animal models is that of getting a paradoxical human model; this is why we have included in this book the results of the use of mediators in neuropsychiatry.

So you will find first an illustration of transmethylation processes in the brain, and then a certain amount of clinical information. The two compartments are not yet communicating, or only with difficulty. Yet we feel that they represent the foundation of any further discussion about methyl donors, whether neurochemical or therapeutic — a discussion that promises to be long and fascinating, and of which this present book is offered as a first outline.

Biochemistry of the Central Nervous System and Behavior

S. S. KETY

It was Hippocrates and his school who challenged the Aristotelian notion that the heart was the seat of emotions and thought, while the major function of the brain was to purify the blood. In a penetrating and poetic passage, the Hippocratic physicians defined the brain as the organ of behavior and reason, a doctrine which has persisted ever since:

And men should know that from nothing else but from the brain come joys, delights, laughter and jests, and sorrows, griefs, despondency and lamentations. And by this, in an especial manner, we acquire wisdom and knowledge, and see and hear and know what are foul and what are fair, what sweet and what unsavory.... And by the same organ we become mad and delirious and fears and terrors assail us, some by night and some by day, and dreams and untimely wanderings and cares that are not suitable and ignorance of present circumstances, desuetude and unskillfulness. All these things we endure from the brain, when it is not healthy, but is more hot, more cold, more moist, or more dry than natural, or when it suffers any other preternatural and unusual affliction.

Although the validity of this doctrine has been generally accepted since their time, the notions of what were the important chemical bases of behavior have evolved with the increasing sophistication of biochemistry and the acquisition of biochemical knowledge.

The time is not yet at handle, when we can speak definitively of the biochemistry of behavior or mental state. There are, however, a few areas where one can see the beginnings of correlations and significant interrelationships. These include consciousness, affective state, and cognitive function, especially as exemplified in their pathologic derangements.

The nature of consciousness has challenged all perceptive individuals including the great philosophers. Although only the most simplistic kind of materialism assumes that its essence is physicochemical, there are obvious correlations between the material world and conscious state, and extending these correlations to include the biochemical functions of the brain seems worthwhile.

Over the past 25 years, measurements have been made by a number of investigators of the rate of energy utilization by the healthy or disordered human brain in terms of oxygen or glucose consumption (7). These studies have indicated that 20 W is the rate of energy utilization necessary in the normal human brain for the maintenance of consciousness and have revealed a strong dependence of consciousness on the rate of energy utilization. Thus, in pathologic states where the oxygen or glucose metabolism would be expected to suffer from a generalized impairment, there is a progressive interference with consciousness. In the relatively mild alterations of consciousness which are seen in senile

Professor of Psychiatry, Harvard Medical School; and Director, Psychiatric Research.
Laboratories, Mailman Research Center, McLean Hospital, Belmont, Mass./USA

dementia, diabetic acidosis, or insulin hypoglycemia, cerebral energy utilization may be depressed by 20%. On the other hand, in surgical anesthesia, insulin or diabetic coma, the reduction may be as much as 40 or 50%.

There are states of altered consciousness, however, in which such a neat correlation with total cerebral energy metabolism does not exist. Normal sleep is one such state and when it was found that the oxygen and energy utilization of the human brain in sleep was not different from that in the waking state (11), evidence was adduced which challenged for the first time the widely accepted assumption made by Sherrington and by Pavlov that sleep was characterized by neuronal inactivity. In mental arithmetic (18), the toxic psychosis produced by LSD (19), and schizophrenia (9), it was also discovered, contrary to expectations, that the energy metabolism of the brain as a whole was not abnormal. One of the conclusions reached, was "that it takes as much oxygen to think an irrational thought as a rational one," but it also became clear that the biochemical processes involved in sleep or in cognition were probably more subtle and complex than could be accounted for merely by the power supply to the brain. The brain, unlike the heart or the liver or kidney, is an organ for computation and communication. In such functions, there is no necessary correlation between the energy utilized and the efficiency of the process or the quality of the output. To differentiate these alterations of consciousness in terms of the cerebral oxygen consumption would be like trying to correlate the nature of a radio program with the power used.

It has become increasingly clear in the past decades that in the function of the synapses of the brain one is more likely to find the basis of higher nervous activity. It would be expected that in these switching mechanisms might lie the differences between sleep and wakefulness, rational and irrational thought, depression and elation. When it became clear that synaptic transmission was largely a chemical process, a basis was provided whereby biochemical changes of even a small and localized nature could have crucial effects upon behavior and mental state.

The few chemical substances which are known to be neurotransmitters and the larger number which are believed to be transmitters and have demonstrable effects on synaptic function are either amines or amino acids. There is little doubt that acetylcholine, the transmitter at the skeletal myoneural junction, is crucial to behavior. Because it is so crucial, however, and so widespread, it can hardly account for much of the remarkable variations in normal behavior.

Three biogenic amines — dopamine, norepinephrine, and serotonin — have assumed prominence over the past decade as likely mediators of affective and motivational states, and less clearly, in cognitive function. This has come about because a number of drugs of fairly recent origin — amphetamine, reserpine, iproniazid, certain tricyclic compounds, and a number of the phenothiazine and butyrophenone classes of compounds, all of which have fairly specific behavioral effects (i.e., euphoria, depression, antidepressant activity, psychotomimetic or antipsychotic) —have clear and rather consistent effects upon one or more of these amines at central synapses.

During the two-and-a-half decades since 1950, there has been a gratifying growth of fundamental research throughout the many disciplines on which psychiatry depends. This has resulted in considerable new knowledge at the psychologic, physiologic, neurochemical and pharmacologic levels about aspects of behavior which appear to represent cardinal features of schizophrenia or the affective disorders: arousal, attention, reward, motivation, exploration, withdrawal, appetitive and aversive behaviors, stereotypy, mood, sleep, and anxiety. New anatomic pathways have been discovered and some already known, like the limbic system, have been richly explored. The physiology, chemistry, and pharmacology of the synapse has become a central focus of neurobiology, new tools have been developed or applied to the CNS — histofluorescence, electron microscopy, cell culture, mathematical, population, and biochemical genetics. In both schizophrenia and manic-depressive psychosis, new evidence has been acquired to indicate the importance of genetic factors in their transmission. New drugs have been discovered with considerable specificity against the cardinal features of schizophrenia and the affective disorders and the accumulation of fundamental knowledge has made possible the productive exploration of their mechanisms of action.

In the current pharmacologic and biochemical approaches to the major psychoses, one sees for the first time the emergence of parsimonious and credible hypotheses regarding the nature of the biological substrates on which these disorders may develop. Recently discovered pathways in the brain characterized by their production of dopamine, norepinephrine, or serotonin have many properties compatible with their involvement in schizophrenic or affective disorder. They have their origin in a relatively small number of neurons concentrated in the brain stem, which, by means of axons with a tendency to branch extravagantly, permeate every other region of the brain presumably exerting some modulation on a wide variety of higher nervous functions (related to mood and motivation). Although it is as yet inconclusive, there is considerable evidence compatible with the involment of one or more of these systems in behavioral states such as attention and arousal, exploration, pleasure and reward, mood, motivation, and even learning. Important convergences appear to be emerging among an increasingly large number of different drugs found to be highly effective in the treatment of major psychoses, and their similar actions on monoamine neurons or their synapses. All of this current activity is notable, not so much for the number of specific hypotheses it has generated capable of explaining the major psychoses — for they are few and far from being validated — but in providing the foundations on which a greater understanding may eventually be built. For the first time, the biological psychiatrist knows some of the questions to ask and what areas remain to be pursued in the attempt eventually to understand the important biological substrates of the psychoses.

Neurochemical Aspects of Mood and Affective Disorder

Pharmacologic observations have provided a wealth of information which implicates the biogenic amines in the neural mechanisms underlying mood and affective disorders in ways which remain to be further explored. The first intimation of this was the finding by Shore and his collaborators (17) that the administration of reserpine to animals was followed by a marked increase in 5-hydroxyindoleacetic acid excretion and a depletion of serotonin in the brain. This was soon followed by the demonstration that this tranquilizing drug, which produced symptoms of severe depression in some patients who received it, interfered with the capacity of axonal endings to store a variety of putative neurotransmitters, including norepinephrine and dopamine as well as serotonin. Shortly thereafter, the excitant properties of iproniazid and its effectiveness as an antidepressant was found to be related to its inhibition of monoamine oxidase, an observation which was readily confirmed by the development of a large number of monoamine oxidase inhibitors which had antidepressant properties. Attention shifted from serotonin to the catecholamines when it was demonstrated that the administration of L-dopa was capable of reversing the depressant effects of reserpine in animals, whereas 5-hydroxytryptophan was relatively ineffective. This was followed by evidence that amphetamine and the tricyclic antidepressant drugs might increase the concentration of norepinephrine at central synapses by favoring its release or blocking its reuptake by the presynaptic terminal. At the present time, there are a large number of observations compatible with the involvement of catecholamines or of serotonin in affective behavior in animals, or mood and affective disorders in man.

Norepinephrine and Affective Disorder

Schildkraut (16), among others, has advanced the hypothesis that some depressions may be associated with an absolute or relative deficiency of catecholamines, particularly norepinephrine, at functionally important receptor sites in the brain, whereas manias may be associated with an excess of such monoamines. This hypothesis was prompted initially by the observation that drugs which elevate mood or relieve depression in man appear to increase the concentration of norepinephrine at central synapses, while drugs which cause severe depression in man deplete norepinephrine in those regions.

The pharmacologic evidence for an increase or decrease of norepinephrine at central synapses is necessarily indirect. An increased synthesis, an increased release from presynaptic endings, and a decreased inactivation of the amine are factors which would be expected to increase its synaptic concentration. Increased synthesis has been inferred from a more rapid turnover of the amine in the brain or by demonstrating an increased activity of the rate-limiting enzyme, tyrosine hydroxylase. Increased release, which has not been demonstrated directly at central synapses, has been inferred from an increased concentration of the O-methylated derivative, normetanephrine, which appears to be formed by catechol-O-methyltransferase acting postsynaptically. Release has been more directly but less physiologically measured from stimulated brain slices in vitro. The major mechanism for synaptic inactivation of norepinephrine is that of its reuptake into the presynaptic ending, which may be indicated by the rate of initial uptake of labeled norepinephrine injected into the cerebrospinal space.

A decreased synaptic concentration is inferred from a decrease in synthesis or an increase in the inactivation of the amine in the presynaptic terminal making less of it available for release. Presynaptic inactivation is accomplished largely by monoamine oxidase and has been estimated by the concentration of deaminated metabolites.

By means of the foregoing techniques, evidence has been accumulated that amphetamine, a well-known euphoriant agent which increases arousal, exploratory activity, and appetitive behavior in animals, inhibits monoamine oxidase, favors the release of norepinephrine and inhibits its reuptake at the presynaptic ending. The monoamine oxidase inhibitors which are effective antidepressants diminish the presynaptic inactivation of norepinephrine and produce a substantial increase in normetanephrine in the brain, thus indicating, the likelihood of an increased concentration of norepinephrine at the synapse. The tricyclic antidepressants appear to block the reuptake of norepinephrine by the presynaptic ending.

Attempts to manipulate norepinephrine levels in the brain sometimes result in behavioral alterations compatible with the hypothesis, but often fail to do so. Many studies have indicated that norepinephrine injected directly into the ventricles produces not the activation which would be expected, but sedation or somnolence. This may be merely a question of appropriate concentration, however, since when norepinephrine is infused into rats at a controlled, low concentration, an increased arousal and exploratory activity has been observed, which changes to sedation when the infused concentration is increased.

Alpha-methyl-paratyrosine, which blocks tyrosine hydroxylase and the synthesis of dopamine and norepinephrine, produces a slight sedation in rats, but hardly as marked a sedation as that achieved with reserpine. In monkeys, such inhibition of catecholamine synthesis has produced a picture not unlike severe retarded depression in man. Amphetamine causes a marked acceleration of self-stimulation in rats which is blocked by alpha-methyl-paratyrosine, suggesting that norepinephrine is important in this appetitive behavior. When alpha-methyl-paratyrosine has been given to man (e.g., in the treatment of hypertension), slight sedation and occasional mild depression have been observed, followed by evidence of hypomania upon its withdrawal. It has terminated mania in relatively few patients in which it has been used, with evidence of a marked reduction in catecholamine metabolism from examination of catecholamine metabolites in urine and CSF.

L-dopa has shown some antidepressant properties in a small percentage of patients with unipolar depression. In a relatively high proportion of patients with bipolar depression, it has induced hypomanic or manic symptoms.

Conclusive evidence has not been obtained for an inadequacy of norepinephrine in the brain of depressed patients nor for an excess in those with mania. A number of studies on the urine show a decrease in the excretion of norepinephrine and normetanephrine in retarded depression and an increase of these metabolites in mania or hypomania. A very small fraction of these urinary amines, however, represent the release or metabolism of norepinephrine in the brain and these findings must be regarded as measures of peripheral, autonomic, and adrenal medullary activity which may parallel changes in mood. On the other hand, there is reason to believe that 3-methoxy-4-hydroxyphenylglycol (MHPG) is a major metabolite of norepinephrine in the brain and that much of the MHPG that appears in the urine is derived from that source. In subhuman primates, it has been shown that approximately 50% of the urinary MHPG originated in the brain. In man, urinary levels of this metabolite parallel closely the mood changes induced by amphetamine, whereas other metabolites of norepinephrine do not. Measurement of urinary MHPG in affective disorders indicates a tendency for levels to be low in depression and elevated in hypomania, although this is not always the case. In some patients with agitated depression, MHPG has been depressed in the urine, even though other norepinephrine metabolites have been elevated, suggesting a closer approximation of MHPG excretion to the central state.

Serotonin and Affective Disorder

The evidence which suggests an involvement of norepinephrine in the pathophysiology and pharmacotherapy of the affective disorders does not exclude an involvement of other putative transmitters of which serotonin is the outstanding contender in terms of present knowledge. Many of the approaches which have been used to advance the norepinephrine hypothesis have been explored, perhaps less exhaustively, for serotonin, with the demonstration of significant effects. The marked sedation in animals and the severe depression in man which can be produced by reserpine were thought at first to result from its depletion of serotonin from axonal endings in the brain. Its reversal by L-dopa rather than by the precursor of serotonin then suggested that it was the norepinephrine depletion which was responsible. Yet that cannot be the whole explanation, since a simple depletion of norepinephrine or of the catecholamines by alpha-methylparatyrosine produces a state in animals and man which is far less severe than that produced by reserpine. Similarly, the evidence that other mood-altering drugs act on norepinephrine applies equally well to serotonin. The monoamine oxidase inhibitors affect levels of serotonin in the brain and presumably its synaptic activity to a greater extent than can be shown for norepinephrine. The tricyclic antidepressants also block the reuptake of serotonin by presynaptic endings and one of them, amitriptyline, appears to have a greater effect on serotonin than on norepinephrine. Repeated electroconvulsive shocks increase serotonin levels in the brain, in fact, as was known before the similar effect on norepinephrine was demonstrated. Thus, the drugs and procedures known to be effective in the treatment of depression appear to increase the levels and presumably the synaptic activity of both norepinephrine and serotonin.

When the levels of serotonin in the brain are manipulated by drugs or the administration of precursors, the effects on behavior in animals and man are qualitatively quite different from those produced by comparable changes in norepinephrine. Elavating serotonin activity by the administration of its immediate precursor, 5-hydroxytryptophan, produces drowsiness, sedation, and lethargy in animals, and blockade of serotonin synthesis by parachlorophenylalanine, which inhibits tryptophan hydroxylase, produces an increase in aggressive behavior, locomotion, and sexual excitation. An agent which appears to destroy serotonergic endings, 5,6-dihydroxytryptamine, produces similar effects. Lesions of the serotonin-containing raphe nuclei of the brain stem deplete serotonin and increase motor activity and arousal, while sleep is inhibited. These effects can be counteracted by 5-hydroxytryptophan. Parachlorophenylalanine in man has been reported to produce anxiety, arousal, and psychotic manifestations reminiscent of effects sometimes reported with L-dopa. In fact, these effects of L-dopa may result from an interference with serotonin action on the basis of competition with the serotonin precursor for transport into the brain or for decarboxylation, and serotonin has been found to be decreased by as much as 60-80% in the brain following large doses of L-dopa. Elevation of serotonin in the brain raises the threshold of pain while its depletion lowers it. In general, these studies do not support a hypothesis that increased serotonin activity would elevate mood and its decrease would depress it. Instead, there is the suggestion of another type of involvement for serotonin at central synapses and in affective states.

Coppen (5), who has advanced the hypothesis that serotonin is crucially involved in the affective disorders, has adduced in its support evidence which is largely clinical. His studies and those of a large number of others can be only briefly summarized.

Several studies of the urinary excretion of 5-hydroxyindoleacetic acid (5HIAA), the major metabolite of serotonin, reveal no consistent pattern in the affective disorders, which is not unexpected. Practically all of the serotonin in the body is present in the gut and the metabolite found in the urine undoubtedly originates there rather than in the brain.

In the CSF, most observers have found a decrease in 5HIAA in depressed patients, although a few do not confirm this. There appear to be some differences between diagnostic subgroups, e.g., unipolar depressives show lower levels than bipolar, psychotic depressives have lower levels than those who are nonpsychotic. Interestingly enough, in mania or hypomania, four studies have shown a decrease in this metabolite, while three showed no change or an increase. The decrease may be more impressive in view of the results obtained with sham mania, where levels of this metabolite in the CSF were increased. Another interesting finding reported by several investigators is that the low levels of this metabolite found in depression appear to persist after recovery. Coppen studied 18 patients with mania, 31 with depression, of whom 8 were restudied after recovery, and compared these with values obtained in 20 hospitalized controls with no psychiatric disturbances. In both the manic and the depressive groups, as well as in the patients in remission, there was a 50% reduction of 5HIAA in the CSF in comparison with the control group. This persistence in both types of affective disturbance and after recovery diminishes the likelihood that irrelevant artifacts are operating, and suggests the presence of some constitutional factor common to both depression and mania.

After probenecid, the accumulation of 5HIAA in CSF was found to be decreased in four studies of depression and two studies of mania.

The effects of precursor administration in the case of serotonin appear to be more consistent than those reported with L-dopa. A few early studies which administered 5-hydroxytriptophan to depressed patients reported no effects or conflicting results. After a report that schizophrenia treated with monoamine oxidase inhibitors and substantial doses of tryptophan showed elevation of mood and loosening of association, two groups evaluated the same combination in patients with depression and both reported significant improvement. There have been five additional studies in which improvement has been reported following large doses of tryptophan with or without monoamine oxidase inhibitors, but four studies report no beneficial effects.

Methysergide, a serotonin antagonist, has been used in mania in nine studies which are evenly divided between improvement and no effect. One investigator found an aggravation of manic symptoms following this drug.

Three groups have examined the brain post mortem for amines and their metabolites, comparing those for suicide victims with those following sudden death with no indication of depression. Although post-mortem studies are fraught with many difficulties and artifacts, the consistent finding emerged from all three studies of a diminution in serotonin or its metabolite in the brains of suicide victims, with no consistent pattern for the catecholamines.

Thus, the evidence for the involvement of serotonin in affective disorder appears to be at least as compelling as that for norepinephrine, and in the case of the clinical studies seems more consistent. Although few of the findings are unanimously confirmed, a tentative conclusion seems permissible that serotonin activity is decreased in the brain in both depression and mania and that the effect appears to be more characteristic of the individual than his particular state of affective disorder or recovery. Carroll (4), after a critical review of the precursor studies, concludes that "the evidence currently available is consistent with the possibility that the rate-limiting hydroxylation of tryptophan may be impaired in depressed patients."

Synthesis

Recent genetic evidence strongly suggests that one or more biochemical disturbances play important roles in a major segment of the serious affective disorders. Research over the past two decades has shown some reasonably reproducible biochemical changes which are associated with affective disorder. Significant alterations undoubtedly occur in electrolyte balance and in corticosteroid secretion. Although their role need not be secondary, no parsimonious concept has as yet emerged relating to the pathogenesis of the affective disorders. On the other hand, fundamental knowledge of the biochemistry and pharmacology of the biogenic amines, their localization and distribution in the brain and their interrelationships with behavior and clinical state, strongly suggests that further study of these areas should elucidate some of the underlying biological processes in depression and mania.

The two major hypotheses that have been advanced, implicating catecholamines and indoleamines, or more specifically, norepinephrine and serotonin, are not mutually exclusive. In fact, by combining them, it may be possible to account for many of the discrepancies. Behavioral studies in animals as well as clinical findings are compatible with the thesis that norepinephrine pathways are involved in mood, with deficiency producing depression and overactivity resulting in euphoria, hypomania, or mania. Similar types of observations with serotonin suggest that its role may be to exert a reciprocal or, more likely, a stabilizing or dampening effect on the synapses the activity of which may be associated with mood. There is rather consistent clinical evidence that serotonin or its activity is reduced in the brain of many patients suffering from affective disorder, whether they be depressed or manic or in remission. It is possible that a deficiency of serotonin at central synapses is an important genetic or constitutional requirement for affective disorder, permitting what might otherwise be normal and adaptive changes in norepinephrine activity and the resultant mood states, to exceed the homeostatic bounds and progress in an undampened fashion to depression or excessive elation. The symptomatic features of depression or mania would thus be attributable to too great or too little norepinephrine activity, while the predisposition to them or the extent to which those changes overrun their adaptive bounds would depend upon a constitutional deficit in serotonin activity. Such a hypothesis, though less simplistic than the two which it combines, does not take into account the actions of other amines and putative transmitters, of which our knowledge at the present time is more limited.

Current Neurochemical Hypotheses of Schizophrenia

Recent Genetic Information

Several studies which have employed adopted individuals and their biological and adoptive families as a means of disentangling genetic from environmental contributions have adduced compelling evidence that genetic factors operate significantly in the transmission of schizophrenia. In one of the studies (8), the finding of a high incidence of schizophrenia in the biological paternal half-siblings of schizophrenics who had been adopted and who did not share with them the same environment, including even the in utero environment or early mothering experience, may represent the most conclusive evidence for genetic factors which has been obtained. The question whether schizophrenia is a homogeneous entity or a phenomenologically similar group of illnesses with different etiologies remains to be answered, although there is some evidence that latent or borderline schizophrenia is genetically related to more severe forms of illness. Two hypotheses relating to the biological substrates of schizophrenia are the foci of considerable current research activity.

Transmethylation

In 1952, Harley Mason speculated (13) that the newly delineated process of biological transmethylation might somehow be involved in schizophrenia. He was impressed that many hallucinogens were methylated substances and that one of these, mescaline, would result from the O-methylation of dopamine at the 3, 4, and 5 positions. Although O-methylation of catecholamines had not yet been described, he postulated that such might occur in the body and that in schizophrenia there might be an accumulation of hallucinogenetic methylated metabolites. He singled out one substance, 3,4-dimethoxyphenylethylamine, as being of special interest since it had been reported to produce catatonic behavior in animals.

This hypothesis received its first test in the administration of niacin or niacinamide in large doses to schizophrenics on the thesis that this substance would be methylated and competitively divert the biological transmethylation process from the production of hallucinogenic substances. The striking therapeutic results which were reported have not been confirmed in a substantial number of controlled trials. Since it has been shown that the administration of niacinamide to animals does not significantly depress brain levels of S-adenosylmethionine, niacinamide is not an effective methyl acceptor in the brain and its administration did not constitute a test of the transmethylation hypothesis. In 1961, the hypothesis was tested in another manner (14), this time by the administration of methionine to schizophrenic patients in conjunction with a monoamine oxidase inhibitor. In approximately one-third of the patients there was a brief exacerbation of psychosis and this phenomenon has been reported in several subsequent studies. In one of these (1), it was found that methionine alone, without monoamine oxidase inhibition, can produce the same manifestation in schizophrenic patients. Although it is difficult to exclude the possibility that the methionine has produced a toxic psychosis superimposed on schizophrenia, normal subjects who receive the same dose of methionine experience no such effect and the results in schizophrenia are compatible with the hypothesis, although alternative possibilities have not been ruled out. The ability of large doses of methionine to elevate S-adenosylmethionine levels in the brain and reports that methionine administration in schizophrenics caused an increase in the excretion of dimethyltryptamine, a well-known hallucinogen, were compatible with the hypothesis, although the latter finding has not been firmly established by unassailable analytical techniques.

In 1962, the occurrence of a substance which produced a pink spot on paper chromatography and was identified as 3,4-dimethoxyphenylethylamine was reported in the urine of schizophrenic patients (6). This was followed by a large number of attempts to replicate it, most of which were successful in demonstrating the pink spot in the urine of schizophrenics and to a considerably smaller extent in normal urine. However, the fact that phenothiazines produced metabolites which also gave a pink spot in a similar position confused the issue. More recently, 3,4-dimethoxyphenylethylamine has been demonstrated by mass fragmentography in the urine of normals and schizophrenics and evidence adduced that it is of exogenous origin. The issue has not yet been resolved.

Evidence for an enzyme, indoleamine-N-methyltransferase in the brain has been obtained (15). That enzyme is capable of methylating tryptamine to dimethyltryptamine, although it does not methylate serotonin.

In 1972, it was shown that 5-methyltetrahydrofolic acid (MTHF) could serve as methyl donor in the N-methylation of dopamine to epinine (10). Banerjee and Snyder (2) have since reported that MTHF can serve as the methyl donor in the methylation of both indoleamines and phenethylamines. They showed that serotonin and dimethylserotonin were actively methylated on the 5-hydroxyl group with MTHF serving as the methyl donor to form the corresponding 5-methoxy compound. One of these, 5-methoxy-N, N-dimethyltryptamine is a very potent hallucinogen, considerably more active than the parent substance bufotenin. The importance of the MTHF pathway in biological transmethylation, however, has recently been seriously challenged.

The Dopamine Hypothesis

Pharmacology offers two approaches to the pathogenesis of a disorder. One is by investigating the mechanism of action of drugs which ameliorate the disorder, the other by examining the actions of drugs which produce or mimic the disorder. In the case of schizophrenia, both of these approaches have been actively pursued.

Since the discovery in 1951 that chlorpromazine was beneficial in the treatment of schizophrenic patients, a number of phenothiazine derivatives have been prepared, many of which have antipsychotic properties. Somewhat later, haloperidol, a butyrophenone not chemically related to the phenothiazines, was also found to be effective in the treatment of schizophrenic patients. As both groups of drugs became widely used, it soon became evident that an important side effect of both was the development of symptoms like those of Parkinson's disease. With the elucidation of the dopamine-containing nigrostriatal pathway and the discovery that dopamine was deficient in the caudate nucleus of patients suffering from Parkinson's disease, an action on that amine became a possibility for explaining the extrapyramidal effects of the antipsychotic drugs. Carlsson and Lindqvist, in 1963, first suggested that these drugs acted by blockade of dopamine receptors (3). They found that chlorpromazine and haloperidol increased the levels of dopamine metabolites in the brain, while promethazine, a phenothiazine drug which is not effective in schizophrenia, did not. With remarkable insight, they speculated that there was an increased release of dopamine in the case of the active antipsychotic drugs brought about by some feedback mechanism in response to a blockade of dopamine receptors. Since that observation, evidence has accumulated to establish that the postulated blockade of dopamine does in fact occur. Several pharmacologic studies have confirmed the increased synthesis and turnover of dopamine in the brain in response to psychoactive congeners and a blockade of dopamine receptors has been demonstrated physiologically by microelectrode recording and biochemically on dopamine-sensitive adenylate cyclase of brain. Matthysse (12) has pointed out that the phenothiazines have a notoriously wide spectrum of effects, from protozoan motility to the pecking behavior of pigeons, and that a criterion for the mechanism involved in antipsychotic activity would be the ability of the effect to discriminate between effective and ineffective phenothiazines and butyrophenones. To a considerable extent, that criterion has been met in the case of dopamine receptor blockade. For two exceptions, thioridazine and clozapine, potent antipsychotic drugs quite free of parkinsonian side effects, Snyder has suggested that their anticholinergic properties may be responsible for preventing the extrapyramidal effects and has demonstrated that these two drugs have the greatest anticholinergic effects of any of the antipsychotic agents.

Although the phenothiazines were at first thought to be useful merely as "tranquilizers" of disturbed behavior, it eventually became apparent that they had rather specific effects on the cardinal features of schizophrenia. Whereas sedative agents like the barbiturates or antianxiety drugs like diazepam were no more effective than placebo in the treatment of schizophrenics, chlorpromazine in large-scale controlled studies produced significant improvement in thought disorder, blunted affect, withdrawal and autistic behavior, all cardinal features as described by Bleuler. The ability of these agents to activate withdrawn patients and bring the characteristic flat affect toward normal is hardly a sedative or tranquilizing effect.

Although LSD commanded considerable interest because of its ability to produce a toxic psychosis thought to resemble schizophrenia, it was known for some time that the psychosis induced by overdosage with amphetamine was much closer to schizophrenia and was often confused with that disorder even by experienced clinicians. Chronic amphetamine toxicity is characterized by a paranoid psychosis with auditory hallucination and stereotyped behavior with little delirium or confusion. Schizophrenic patients have been reported readily able to recognize an LSD psychosis as different from their usual symptoms, but unable to differentiate an amphetamine psychosis. Furthermore, amphetamine, methylphenidate and L-dopa can precipitate active schizophrenia in schizophrenic patients during remission.

There is evidence to suggest that the psychosis of amphetamine is mediated through its release and potentiation of dopamine at its receptors in the brain. The stereotyped behavior induced in animals by amphetamine can be prevented by lesions of the dopamine pathways. Dopamine itself or apomorphine, which is known to stimulate dopamine receptors, will produce stereotypy when injected into the brain. Behavioral and biochemical observations suggest that d-amphetamine is considerably more potent in potentiating norepinephrine than is l-amphetamine, while both isomers of amphetamine have equivalent actions on dopamine synapses. Whereas in man the d-form is approximately five times as potent as the l-form of amphetamine in its arousal and euphoriant effects, suggesting that these are mediated by noradrenergic pathways, the two forms of amphetamine are almost equipotent in producing psychosis. Further suggesting that the amphetamine psychosis is the result of an activation of dopamine synapses is the observation that chlorpromazine and haloperidol are very effective in preventing or aborting the psychosis.

There is thus a remarkable convergence upon dopamine synapses in the brain on the part of several types of agents. The drug which produces a psychosis most closely resembling schizophrenia appears to act by potentiating dopamine at its synapses in the brain, whereas a large number of drugs in two distinct chemical classes which have in common their ability to block dopamine receptors also have quite specific effects upon the cardinal features of schizophrenia. This had led to the hypothesis that an overactivity of dopamine synapses may play a crucial role in the pathogenesis of schizophrenia.

A Dopamine: Norepinephrine Imbalance?

Although a hyperactivity of dopamine could explain certain features of schizophrenia such as stereotyped behavior, paranoid delusions and auditory hallucinations, it would hardly in itself account for other features of schizophrenia, such as anhedonia, withdrawal and autism, flatness of affect. These manifestations appear to involve behavioral components which in animals have been related to the activity of norepinephrine. Thus, there is considerable indirect evidence that norepinephrine pathways are involved in appetitive or reward behavior, in exploratory activity, and in elevated mood. It is possible then that the corresponding manifestations of anhedonia, withdrawal, and flatness of affect represent an insufficiency of norepinephrine at its synapses in schizophrenia. A parsimonious mechanism is hypothesized which could produce both the increase in dopamine activity and the decrease in norepinephrine activity which the more complete explanation of the symptoms would require. That mechanism centers on the enzyme dopamine-beta-hydroxylase, which converts dopamine to norepinephrine at the adrenergic endings in the brain. A throttling of that enzyme could conceivably result in the release of dopamine at the expense of norepinephrine.

There are other findings in schizophrenia related to serotonin, histamine and acetylcholine, although they are few and their significance remains to be explored. A reciprocal relationship between cholinergic and catecholamine pathways is suggested by the ability of physostigmine to prevent the acute exacerbation of psychosis which can be precipitated in schizophrenia during remission by methylphenidate. An insensitivity of schizophrenics to histamine as evidenced by a diminution in the wheal produced by that amine has not yet been explained.

One of the most striking findings to emerge from biological research in schizophrenia is the finding by Wyatt and collaborators (20) of a significantly diminished level of monoamine oxidase in the platelets of schizophrenic patients. That finding has assumed greater significance in conjunction with their observation that the diminution in that enzyme holds true even for the nonschizophrenic monozygotic twin in pairs discordant for schizophrenia. This finding, which should be free of the influence of drugs, institutionalization and other nondisease variables which often confuse the interpretation, suggests that the reduced levels of that enzyme in schizophrenia is a genetic characteristic. If it were also to involve that enzyme in the brain, its relevance to the behavioral and mental alterations would be quite clear. The results obtained in five studies which have assayed monoamine oxidase in the brain of schizophrenics are not consistent.

Although biochemistry is far from supplying an adequate explanation of behavior, mood, cognition, and their disturbances in the major psychoses, it is apparent that considerable progress has been made in recent years. The relevance of certain areas of fundamental research, notably the synaptic functions of the biogenic amines and their relationships to behavior have been more clearly established and a modicum of cautious optimisms seems warranted.

16

References

1. Antun, F.T., Burnett, G.B., Cooper, A.J., Daly, R.J., Smythies, J.R., Zealley, A.K.: The effects of l-methionine (without MAOI) in schizophrenia. J. Psychiat. Res. **8**, 63 (1971)
2. Banerjee, S.R., Snyder, S.H.: Methyltetrahydrofolic acid mediates N- and O-methylation of biogenic amines. Science **182**, 74 (1973)
3. Carlsson, A., Lindqvist, M.: Effect of chlorpromazine or haloperidol on formation of 3-methoxytryptamine and normetanephrine in mouse brain. Acta Pharm. toxicol. **20**, 140 (1963)
4. Carroll, B.J.: Monoamine precursors in the treatment of depression. Clin. Pharm. Ther. **12**, 743 (1973)
5. Coppen, A.: Indoleamines and affective disorders. In: Symposium on Biochemical and Pharmacological Aspects of Affective Disorders. Schildkraut, J.J. (ed.) J. Psychiat. Res. **9**, 163 (1972)
6. Friedhoff, A.J., Winkle, E. van: Characteristics of an amine found in urine of schizophrenic patients. J. nerv. ment. Dis. **135**, 550 (1962)
7. Kety, S.S.: The general metabolism of the brain in vivo. In: Metabolism of the Nervous System. Richter, D. (ed.) London: Pergamon 1957, pp. 221-237
8. Kety, S.S., Rosenthal, D., Wender, P.H., Schulsinger, F., Jacobsen, B.: Mental illness in the biological and adoptive families of adopted individuals who have become schizophrenic. In: Genetics and Psychopathology. Fieve, R., Brill, H., Johns Hopkins, Rosenthal, D. (eds.) Baltimore: 1974
9. Kety, S.S., Woodford, R.B., Harmel, M.H., Freyhan, F.A., Appel, K.E., Schmidt, C.F.: Cerebral blood flow and metabolism in schizophrenia. The effects of barbiturate semi-narcosis, insulin coma and electroshock. Amer. J. Psychiat. **104**, 765 (1948)
10. Laduron, P.: N-methylation of dopamine to epinine in brain tissue using N-methyltetrahydrofolic acid as the methyl donor. Nature (New Biol.) **238**, 212 (1972)
11. Mangold, R., Sokoloff, L., Conner, E., Kleinerman, J., Therman, P.G., Kety, S.S.: The effects of sleep and lack of sleep on the cerebral circulation and metabolism of normal young men. J. clin. Invest. **34**, 1092 (1955)
12. Matthysse, S.: Antipsychotic drug actions: a clue to the neuropathology of schizophrenia? Fed. Proc. **32**, 200 (1973)
13. Osmond, H., Smythies, J.: Schizophrenia: a new approach. J. Ment. Sci. **98**, 309 (1952)
14. Pollin, W., Cardon, P.V., Kety, S.S.: Effects of amino acid feedings in schizophrenic patients treated with iproniazid. Science **133**, 104 (1961)
15. Saavedra, J.M., Coyle, J.T., Axelrod, J.: The distribution and properties of the non-specific N-methyltransferase in brain. J. Neurochem. **20**, 743 (1973)
16. Schildkraut, J.J.: Neuropsychopharmacology and the Affective Disorders. Boston: Little, Brown and Co., 1970
17. Shore, P.A., Pletscher, A., Tomich, E.G., Carlsson, A., Kuntzman, R., Brodie, B.B.: Role of brain serotonin in reserpine action. Ann. N.Y. Acad. Sci. **66**, 609 (1957)
18. Sokoloff, L., Mangold, R., Wechsler, R.L., Kennedy, C., Kety, S.S.: The effect of mental arithmetic on cerebral circulation and metabolism. J. clin. Invest. **34**, 1101 (1955)
19. Sokoloff, L., Perlin, S., Kornetsky, C., Kety, S.S.: The effects of d-lysergic acid diethylamide on cerebral circulation and over-all metabolism. Ann. N.Y. Acad. Sci. **66**, 468 (1957)
20. Wyatt, R.J., Murphy, D.L., Belmaker, R., Cohen, S., Donnelly, C.H., Pollin, W.: Reduced monoamine oxidase activity in platelets: a possible genetic marker for vulnerability to schizophrenia. Science **179**, 916 (1973)

Transmethylating Enzymes

J. M. SAAVEDRA

The transmethylases form a large subclass (EC 2.1.1) of the transferases, and specifically of the transferase enzymes that transfer one-carbon groups (25). They form a large group, of 41 different enzymes, with a wide variety of functions and characteristic requirements. They catalyse the transfer of a methyl group from one compound, or "donor," to another compound, or "acceptor." In most cases, S-adenosyl-L-methionine was found to be the required methyl donor.

The transmethylases are involved in the metabolism of many compounds of great biological interest, including porphyrins, fatty acids, phospholipids, polysaccharides, proteins, nucleic acids, and biogenic amines. A great number of drugs and foreign compounds, such as amphetamines, tricyclic antidepressants, phenols, normorphine and nornicotine, can also be methylated and thus metabolized by a nonspecific transmethylating enzyme.

We will confine our description to only a few of these transmethylases, ones that play important roles in the regulation of the nervous system activity, that may be associated with the pathophysiology of some mental illnesses, and that are known to metabolize endogenous and exogenous compounds of major importance for neuropsychiatry.

The most important characteristic of this group of transmethylases to be considered here is that they affect the synthesis and inactivation of biogenic amines, especially neurotransmitters, in mammalian systems.

Depending on the enzyme, transmethylases can transfer the methyl group from S-adenosyl-L-methionine to either oxygen or nitrogen to form a methylated product and homocysteine. These enzymes are essential for the metabolism of norepinephrine, dopamine, epinephrine, serotonin, and histamine. They have a selective substrate specificity and distribution in tissues, and their activity is finely regulated by a variety of neuroendocrine mechanisms (Table 1).

We shall consider here:

1) 2.1.1.4   Acetylserotonin methyltransferase (also hydroxyindole-O-methyltransferase, HIOMT) (12)
2) 2.1.1.6   Catechol methyltransferase (also catechol-O-methyltransferase, COMT) (9)
3) 2.1.1.8   Histamine methyltransferase (HMT) (17)
4) 2.1.1.28 Norepinephrine N-methyltransferase (also phenylethanolamine-N-methyltransferase, PNMT) (2)
5) Nonspecific methyltransferase (3)
6) The use of transmethylating enzymes as analytical tools in neurochemistry.

National Institute of Mental Health, Washington D.C./USA.

Table 1. General characteristics of transmethylating enzymes

Enzyme	Localization	Physiologic substrates	Function
Catechol-O-methyl-transferase (COMT)	Widespread. Peripheral nerves, CNS, liver, red blood cells	Norepinephrine Dopamine Epinephrine	Physiologic inactivation of catecholamines
Hydroxyindole-O--methyltransferase (HIOMT)	Exclusively pineal gland	N-acetyl-serotonin	Formation of melatonin
Histamine-N-methyl-transferase (HMT)	Widespread. Brain	Histamine	Metabolism of histamine
Phenylethanolamine -N-methyltransferase (PNMT)	Adrenal gland, brain, heart	Norepinephrine and other phenylethanol-amines	Formation of epinephrine
Nonspecific N-methyl transferase	Widespread. Lung	Wide variety of endogenous and exogenous compounds	Unknown

1) 2.1.1.4 Acetylserotonin Methyltransferase (Hydroxyindole-O-Methyl-transferase, HIOMT)

Localization. HIOMT is an enzyme characteristic of the pineal gland (52). It has been found in the pineal glands of all vertebrate species studied (8, 13, 36). In mammals, HIOMT is found only in the soluble supernatant of the pineal gland (11, 12). While in mammals HIOMT activity is found almost exclusively in the pineal gland, the enzyme has also been detected in the brain of amphibians and other nonmammalian species (36).

Substrates. N-acetylserotonin has been found to be the best substrate for HIOMT. This property has recently been utilized to develop a sensitive enzymatic-isotopic assay for N-acetylserotonin (prior to N-acetylation) (18, 41). Other 5-hydroxyindoles, such as serotonin, N-N-dimethylserotonin (bufotenin) and 5-hydroxytryptophol, are also O-methylated by this enzyme, but to a lesser degree (12). When the hydroxy group is substituted on other positions on the indole nucleus, no O-methylation is detected (12).

Cofactors, Characteristics, Inhibitors. All methyltransferases are inhibited by relatively low doses of p-chloromercuribenzoate, indicating the presence of an essential sulfhydryl group. Other potent HIOMT inhibitors include several N-acetyltryptamine derivatives (16), and S-adenosyl-homocysteine (23), by a competitive mechanism.

Control of Activity. Changes in physiologic conditions and environmental lighting can modify HIOMT activity in rats. In the rat pineal, the enzyme activity increases when the animals are kept in constant darkness (53). When the tracts connecting the brain with the eyes, or the sympathetic nerves innervating the gland, are cut, this effect of light disappears, indicating its indirect origin (51).

Measurement of Activity. HIOMT activity is measured by the incubation of pineal supernatant with N-acetylserotonin, and ^{14}C-methyl-S-adenosylmethionine (12). The enzymatically formed ^{14}C-melatonin is separated by extraction into toluene or chloroform at alkaline pH, from the radioactive S-adenosylmethionine.

Function and Biological Importance. HIOMT catalyses the last step in the production of the active pineal principle, the hormone melatonin. Melatonin is indeed a very active physiologic compound. It lightens the skin of amphibians when administered in minute amounts; this effect has been utilized as a biological assay for melatonin. Although the possible effects of melatonin in the regulation of neuroendocrine function are at best only partially understood, it is clear that when administered parenterally, this compound produces antigonadotrophic effects. Thus, in the rat, melatonin injections inhibit the estrous phase of the estrous cycle.

HIOMT has been found in pineal tumors in man and has been used as a marker for identifying tumors of pineal origin (55).

2) 2.1.1.6 Catechol Methyltransferase (also Catechol-O-Methyltransferase, COMT)

Localization. COMT is widely distributed in nature, not only in animal tissues (9), but also in plants (26). In mammals, it is present in high amounts in the liver, and it has also been detected in human red blood cells (5) and in human platelets (Saavedra and Axelrod, unpublished observations). COMT activity has also been detected throughout the CNS, and in peripheral nerves (7). Most of the COMT activity is found in the soluble fraction of the cell (9); however, the existence of membrane-bound COMT has also been reported (30) and in some cases it forms a substantial portion of the whole enzyme activity (Saavedra and Axelrod, unpublished observations).

Substrates. The enzyme O-methylates all catechols, regardless of the substituent on the aromatic nucleus. Normally occurring compounds, such as norepinephrine, epinephrine, dopamine and its metabolites, and dopa, are O-methylated. Other compounds, such as ascorbic acid and 2-hydroxyestradiol, are also O-methylated. Enzymatic O-methylation occurs predominantly on the meta position. Figure 1 shows the metabolism of the neurotransmitter norepinephrine by COMT and monoamine oxidase.

The ability of COMT to O-methylate catecholamines specifically has recently been used in the development of an assay for norepinephrine and dopamine in tissues (22).

Cofactors, Characteristics, Inhibitors. Unlike other transmethylases, COMT requires the presence of Mg^{++} or other divalent cations. The enzyme activity is inhibited by p-chloromercuribenzoate, which suggests the presence of an essential sulfhydryl group. COMT

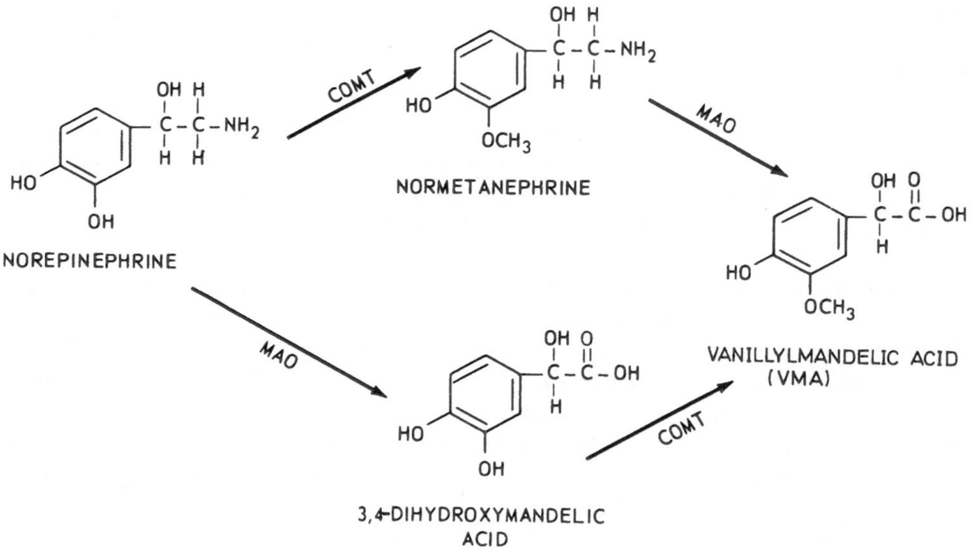

Fig. 1. Metabolism of norepinephrine by COMT and monoamine oxidase

can also be inhibited by a number of different compounds, for example polyphenols and catechols such as pyrogallol (6), and tropolones (15).

The enzyme does not seem to be homogeneous when submitted to purification procedures; the presence of COMT isoenzymes has been reported for a number of rat tissues (10). The physiologic significance of these isoenzymes is at present unknown.

Control of Activity. Little is known about the physiologic mechanisms for control of COMT activity. Soluble and membrane-bound COMT seem to have different regulatory mechanisms; cold stress and benzopirene can increase the microsomal COMT, whereas the soluble form of the enzyme does not respond to these treatments.

Steroids may have a physiologic role in the control of enzyme activity; pregnancy results in a twofold increase in COMT activity in the rat uterus (54); hypophysectomy has been shown to reduce COMT activity in rat liver (33).

Measurement of Activity. COMT activity can be measured by the use of norepinephrine as substrate, and ^{14}C-methyl-S-adenosyl-L-methionine (9). The product, normetanephrine, is extracted into a mixture of toluene and isoamylalcohol, and the radioactivity is counted. Recently, the use of ^{3}H-methyl-S-adenosyl-L-methionine of high specific activity provided the assay with a 100-fold increase in sensitivity (Saavedra, unpublished observations). With this modified assay, COMT can now be measured in as little as 50 μg of brain tissue. COMT activity has been shown to be unevenly distributed in discrete rat brain nuclei (Saavedra, unpublished observations).

Function and Biological Importance. COMT has an important role in the metabolic inactivation of tissue and circulating catecholamines, (including the neurotransmitter nor-

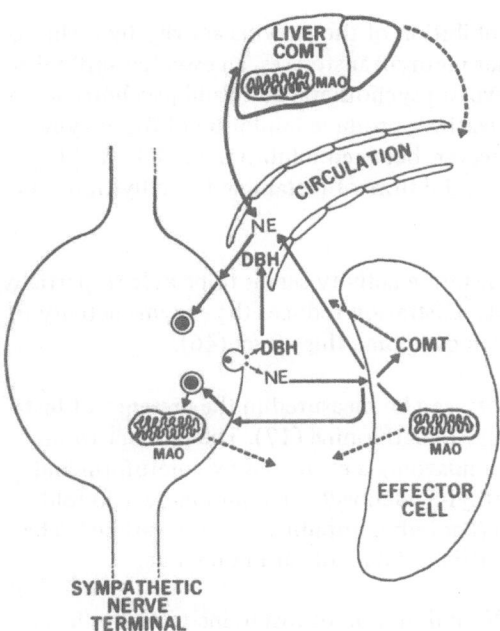

Fig. 2. Sites of action of catechol-O-methyltransferase (COMT) in the metabolism of the neurotransmitter norepinephrine (NE). MAO = mono-amine oxidase; DBH = dopamine-beta-hydroxylase

epinephrine and the hormone epinephrine). Figure 2 shows two of the known sites of action for COMT, the liver, which metabolizes catecholamines coming from the circulation, and effector cells, in which COMT is also present.

Several O-methylated metabolites of catecholamines, among them normetanephrine, metanephrine 3-methoxytyramine, and 3-methoxy-4-hydroxymandelic acid, are excreted in the urine. Their amounts offer an indirect estimate of catecholamine metabolism in tissues.

COMT also metabolizes a portion of the neurotransmitter norepinephrine that is released from the nerve endings. Membrane-bound COMT may be closely related to the adrenergic receptor (24).

3) 2.1.1.8 Histamine Methyltransferase (HMT)

Localization. Histamine N-methyltransferase is widely distributed in mammalian tissues, including the brain, which possesses the highest activity (17, 47), and human red blood cells (5). In mammalian tissues, the enzyme activity is mainly confined to the soluble supernatant fraction of the cell. The neurohypophysis and the hypothalamus contain very high levels of histamine N-methyltransferase (7). In the brain, the highest activity is localized in the hypothalamus (47).

Substrates. The enzyme is highly specific for histamine (17); other imidazole derivatives are not substrates for histamine N-methyltransferase.

Cofactors, Characteristics, Inhibitors. The inhibition of the enzyme activity by p-chloro-mercuribenzoate indicates that, as with other methyltransferases, an essential sulfhydryl group is involved in the enzyme activity. Several psychotherapeutic and psychotoxic drugs, such as chlorpromazine and LSD derivatives, produce inhibition of the enzyme activity. No correlation has been found, however, between inhibition of HMT and behavioral effects. Antihistamine drugs are also inhibitors of histamine N-methyltransferase (45).

Control of Activity. Histamine N-methyltransferase activity seems to be at least partially controlled by testosterone and other steroids. Castration reduces the enzyme activity in male rats, and testosterone administration can overcome this effect (46).

Measurement of Activity. The enzyme activity can be measured in the presence of histamine, as substrate, and ^{14}C-methyl-S-adenosyl-L-methionine (17). The product formed, methylhistamine (1-methyl [β-aminoethyl] imidazole) is extracted by chloroform and the radioactivity measured. By using ^{3}H-methyl-S-adenosyl-L-methionine, a 100-fold increase in sensitivity was recently obtained (Saavedra, unpublished observations). The enzyme activity can now be measured in as little as 50 μg of rat brain tissue.

Functions and Biological Importance. The N-methylation of histamine to methylhistamine is the major pathway for histamine catabolism in mammals.

Histamine has recently been shown to be concentrated in the median eminence of the hypothalamus (19). By the use of the new sensitive method for histamine N-methyltransferase, this enzyme has also been shown to be concentrated in the hypothalamus.

4) 2.1.1.28 Norepinephrine N-Methyltransferase (also Phenylethanolamine N-Methyltransferase, PNMT)

Localization. PNMT N-methylates norepinephrine to form epinephrine; it is therefore involved in the last step in the metabolism of catecholamines.

In mammals, it is found almost exclusively in the soluble supernatant fraction of the adrenal medulla (2). Small amounts of PNMT activity are present in the mammalian heart and brain (4), as well as in the superior cervical ganglia and organ of Zuckerkandl of the new born rat. With the use of immunohistofluorescent techniques (29) and a sensitive microassay (43), PNMT has recently been detected in specific nuclei of the rat brain stem and hypothalamus, including the median eminence. The enzyme activity is not homogeneously distributed among catecholamine-rich nuclei in the brain stem; C1 and C2 areas show the highest activity. Specific adrenergic tracts in the brain, therefore, seem to occur (43).

PNMT is widely distributed in the toad; it is present in the heart, brain, adrenal gland, parotid gland (34), and also in the sciatic nerve, where it is transported down the axons by a slow energy-dependent mechanism (49).

Substrates. Not only norepinephrine, but also primary and secondary β-hydroxylated phenylethylamines (normetanephrine, metanephrine, epinephrine, phenylethanolamine, and octopamine) are preferentially N-methylated by PNMT (4). Beef adrenal PNMT can also methylate phenylethylamines, although to a lesser extent (28). However, p-chloro-phenylethylamine is a relatively good substrate for the enzyme (Saavedra, unpublished observations).

Cofactors, Characteristics, Inhibitors. High phosphate concentrations, above 0.05 M, produce inhibition of the enzyme activity. As with other methyltransferases, PNMT activity is also inhibited by p-chloromercuribenzoate, indicating the presence of an essential sulfhydryl group. There is pronounced substrate inhibition of PNMT activity with high concentrations of phenylethanolamine, norepinephrine, epinephrine, and also β-phenylethylamine (27). The monoamine oxidase inhibitor tranylcypromine and related compounds also inhibit PNMT activity in vitro, but not in vivo (31).

Control of Activity. PNMT activity in the adrenal gland is controlled by adrenal gluco-corticoids (50).

After hypophysectomy, adrenal PNMT activity falls 80% in the first week (50). It was found that glucocorticoids stimulate the formation of new molecules of PNMT (4). The fall in PNMT activity after hypophysectomy can be reversed with either dexame-tasone treatment or ACTH treatment.

On the other hand, the administration of glucocorticoids does not increase PNMT activity in normal animals. In newborn rats, however, glucocorticoids stimulate superior cervical ganglion PNMT, when injected during the first postnatal days (21).

PNMT activity can also be influenced by nerve impulses (48). Prolonged nerve stimu-lation after injection of reserpine or 6-hydroxydopamine causes an elevation of adrenal PNMT in the rat.

The genetic control mechanism for PNMT activity has recently been studied in mice (20). Analysis of adrenal PNMT activity in sublines of the Balb/c inbred mouse strain suggests that the enzyme levels are controlled by a single genetic locus (20).

Measurement of Activity. The enzyme activity is measured by determining the ^{14}C-me-thylphenylethanolamine formed after incubation with phenylethanolamine and ^{14}C-methyl-S-adenosyl-L-methionine. The radioactive product is separated from the methyl donor by extraction into a mixture of toluene-isoamyl-alcohol at pH 10. The enzyme activity is then measured in an aliquot of the organic extract by the addition of phos-phor.

The use of ^{3}H-methyl-S-adenosyl-L-methionine of high specific activity as methyl donor, and the introduction of a drying procedure to eliminate a volatile contaminant, increased the sensitivity of the assay 50-fold (43), and allowed the determination of PNMT activity in nuclei from rat brain weighing less than 1 mg.

Functions and Biological Importance. PNMT is involved in the last step of catecholamine synthesis, namely the formation of the hormone epinephrine, in the adrenal medulla.

The recent report of the presence of PNMT in specific regions and nuclei of the rat brain suggests that this enzyme may have a role in the CNS of mammals. The probable production of epinephrine in the median eminence indicates that this biogenic amine may be related to the control of important neuroendocrine mechanisms.

5) Nonspecific N-Methyltransferase

Localization. Nonspecific N-methyltransferase was first described by Axelrod in the rabbit (3). The enzyme activity was highly localized in the lung of several mammalian species, including man (4). The activity was confined to the tissue supernatant.

Nonspecific N-methyltransferase activity has recently been detected in mammalian brain, including that of man (42). In the brain, maximum activity was found in cortical areas of both rat and man (42). Human blood platelets also contain nonspecific N-methyltransferase activity (56).

Substrates. The nonspecific N-methyltransferase from rabbit lung can N-methylate a wide variety of compounds, including biogenic amines (serotonin, tryptamine, N-methyltryptamine, tyramine, β-phenylethylamine, norepinephrine), drugs (desmethylimipramine, amphetamine, normorphine, nornicotine), and other foreign amines (aniline).

In the brain, the enzyme is also relatively nonspecific; it methylates tryptamine as well as phenylethylamine derivatives, but it cannot methylate serotonin when S-adenosyl-L-methionine is used as methyl donor (42). The presence of an enzyme in rat brain that uses methyltetrahydrofolic acid as methyl donor has recently been reported (32, 14). This enzyme seems to be different from that requiring S-adenosylmethionine.

Cofactors, Characteristics, Inhibitors. Antipsychotic drugs like chlorpromazine and imipramine have been shown to inhibit nonspecific N-methyltransferase activity in concentrations that are in the therapeutic range (3).

The presence of potent and normally occurring dialyzable inhibitors for N-methyltransferase activity in brain and human platelets has recently been described (38, 42, 56).

Control of Activity. Little is known about the control mechanisms for nonspecific N-methyltransferase activity. The presence of potent endogenous inhibitors suggests that this enzyme is tightly controlled under normal circumstances. The nature of these inhibitors is at present unknown.

Measurement of Activity. Nonspecific N-methyltransferase activity can be measured in dialyzed tissue supernatants (42). Several substrates (tryptamine, β-phenylethylamine, or N-methyltryptamine) are used. The N-methylated products obtained in the reaction are separated from S-adenosyl-L-methionine by solvent extraction, and the radioactivity counted.

Function and Biological Importance. Nonspecific N-methyltransferase can N-methylate normally occurring tryptamine (38) to form N-methyltryptamine, and this compound can accept a second methyl group to form the psychotomimetic compound dimethyltryptamine (DMT) (1). Figure 3 shows the formation of DMT from tryptamine by the N-methylating enzyme.

The in vivo formation of DMT by the rat brain, after administration of tryptamine, has recently been demonstrated (38). Figure 4 shows the identification of radioactive N-methyl- and N, N-dimethyltryptamine (DMT) formed by the rat brain after administration of the radioactive precursor, tryptamine.

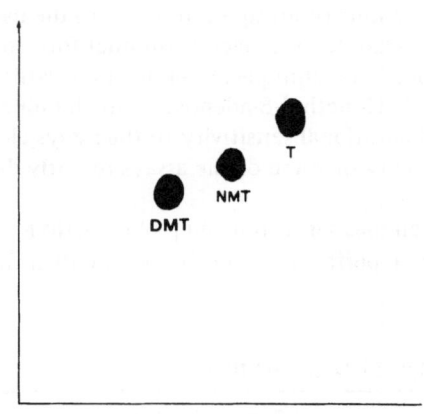

DIMETHYLTRYPTAMINE

Fig. 3. Formation of psychotomimetic tryptamines

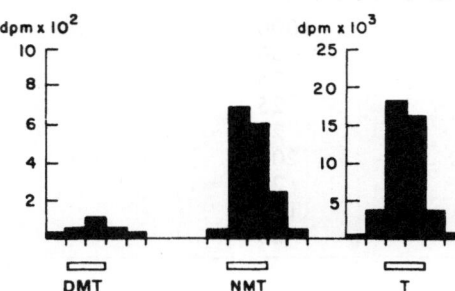

Fig. 4. Chromatographic identification of the formation of N-methylated metabolites of tryptamine in the intact rat brain. The solvent systems used were n-butanol: acetic acid: water (12:3:5) for the first dimension (vertical in the figure) and isopropanol: 10% ammonium hydroxide: water (200:10:20) for the second dimension (horizontal in the figure) DMT = dimethyltryptamine; NMT = methyltryptamine; T = tryptamine; dpm = disintegrations per minute

The presence of an enzyme in human tissues that could produce psychotomimetic metabolites is of potential interest to psychiatry, especially since the ability of human platelets to form DMT was higher in schizophrenic patients than in normal matched controls (56). This difference in enzyme activity was due to a reduced dialyzable enzyme inhibitor in schizophrenic patients.

6) Transmethylating Enzymes as Analytical Tools in Neurochemistry

Over the last years, a number of sensitive enzymatic isotopic micromethods to measure biogenic amines in tissues and small amounts of body fluids have been developed. These methods are based on the incubation of the biological material with specific methyltransferase enzymes (9, 12, 2, 3, 17), together with S-adenosyl;L-methionine. By the use of labeled S-adenosyl;L-methionine as a donor of radioactive methyl groups, radioactive N- or O-methyl derivatives of the biogenic amines are formed. These derivatives are then separated by means of selective solvent extraction procedures, and the radioactivity is counted.

The two major considerations in setting up an assay for biogenic amines are specificity of the assay and sensitivity. Specificity of the assays was obtained by: (a) the use of methylating enzymes with specificity for a given amine or group of amines, (b) the use of solvents of different degrees of polarity to separate the radioactive product formed in the reaction, and (c) the application of evaporation techniques to eliminate volatile radioactive contaminating substances. The use of ^{3}H-methyl-S-adenosyl-L-methionine with high specific activity has recently provided additional sensitivity to the assays already available. Table 2 shows the current sensitivity of some of the assays recently developed.

In some instances, such as with the biogenic amines serotonin and β-phenylethylamine, direct methylation can not conveniently be performed. The amines are then first

Table 2. Enzymatic-isotopic assays using transmethylating enzymes

Amine	Enzyme	Sensitivity (pg)
Tryptamine	NMT	1000
N-acetylserotonin	HIOMT	50
Serotonin	NAT-HIOMT	25
Phenylethanolamine	PNMT	25
Octopamine	PNMT	25
β-Phenylethylamine	DBH-PNMT	200
Histamine	HMT	25
Dopamine	COMT	100
Norepinephrine	COMT	25

NMT: Nonspecific methyltransferase
HIOMT: Hydroxyindole-O-methyltransferase
NAT: N-acetyltransferase (EC 2.3.1.5) from rat liver
PNMT: Phenylethanolamine N-methyltransferase
DBH: Dopamine-β-hydroxylase from bovine adrenal
HMT: Histamine methyltransferase
COMT: Catechol-O-methyltransferase

$$\text{Serotonin} \xrightarrow[\text{Acetyl CoA}]{\text{NAT}} \text{N-acetylserotonin} \xrightarrow[^3\text{H-SAMe}]{\text{HIOMT}} \text{Melatonin}$$

Fig. 5. Enzymatic-isotopic assay for serotonin. After extraction from the tissues, serotonin is first N-acetylated to form N-acetylserotonin by the enzyme N-acetyltransferase from rat liver, in the presence of acetyl coenzyme A. The enzymatically formed N-acetylserotonin is then converted to radioactive melatonin by O-methylation by HIOMT, in the presence of ^{3}H-methyl-S-adenosyl-L-methionine (^{3}H-SAMe)

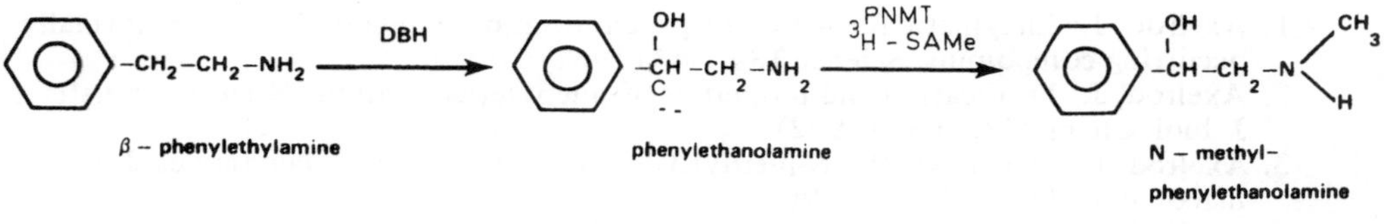

Fig. 6. Enzymatic-isotopic assay for β-phenylethylamine. After extraction from the tissues, β-phenylethylamine is converted into phenylethanolamine by the enzyme dopamine-β-hydroxylase from bovine adrenal gland. The phenylethanolamine is then N-methylated by PNMT in the presence of ^{3}H-methyl-S-adenosyl-L-methionine (^{3}H-SAMe). The radioactive N-methylphenylethanolamine formed is separated from the methyl donor by solvent extraction

enzymatically converted into derivatives with a higher affinity for methyltransferase enzymes, and the methylation is then performed (see Figure 5 and 6).

These assays have been applied to a number of animal and clinical studies. It was possible to detect and measure for the first time octopamine, β-phenylethylamine, phenylethanolamine, and tryptamine in the brain of mammals (39, 40, 37). Furthermore, some of these assays have been applied to the measurement of dopamine, norepinephrine, histamine, and serotonin in discrete nuclei of the rat brain, weighing less than 0.5 mg. The results obtained allowed for the first time the exact quantitation of biogenic amines in regions like the median eminence (Table 3). These results suggest that the biogenic amines studied could be related to the regulation of neuroendocrine function.

Some of the methods are currently applied to the study of biogenic amine metabolism in plasma and other tissues of normal volunteers and neuropsychiatric patients.

Table 3. Biogenic amines in the rat median eminence

	Concentration[a] ng/mg protein
Serotonin	15.3
Histamine	17.8
Norepinephrine	29.5
Dopamine	65.)

[a] Data from (19, 35, 44).

In conclusion, the transmethylating enzymes here described are involved in the metabolism and regulation of a great number of compounds of great biological significance. A study of these enzymes will help to understand the basic mechanisms of nervous function. The enzymes can also be used as tools for the study of specific substrates of physiologic importance.

References

1. Axelrod, J.: Enzymatic formation of psychotomimetic metabolites from normally occurring compounds. Science **134**, 343 (1961)
2. Axelrod, J.: Purification and properties of phenylethanolamine N-methyltransferase. J. biol. Chem **237**, 1657 (1962)
3. Axelrod, J.: The enzymatic N-methylation of serotonin and other amines. J. Pharmacol. exp. Ther. **138**, 28 (1962)
4. Axelrod, J.: Methyltransferase enzymes in the metabolism of physiologically active compounds and drugs. In: Handbook of Experimental Pharmacology, New Series. Vol. XXVIII/2 Brodie, B.B., Gilette, J. (eds.). Berlin-Heidelberg-New York: Springer 1971
5. Axelrod, J., Cohn, C.K.: Methyltransferase enzymes in red blood cells. J. Pharmacol. exp. Ther. **176**, 650 (1971)
6. Axelrod, J., La Roche, M.J.: Inhibition of O-methylation of epinephrine and norepinephrine in vitro and in vivo. Science **130**, 800 (1959)
7. Axelrod, J., MacLean, P.D., Albers, R.W. and Weissbach, H.: Regional distribution of methyltransferase enzymes in the nervous system and glandular tissues. In: Regional Neurochemistry pp. 307-311 Kety, S.S., Elkes, J. (eds.). Oxford: Pergamon, 1961
8. Axelrod, J., Quay, W.B., Baker, P.C.: Enzymatic synthesis of the skin-lightening agent melatonin in amphibians. Nature (Lond.) **208**, 383 (1965)
9. Axelrod, J., Tomchick, R.: Enzymatic O-methylation of epinephrine and other catechols. J. biol. Chem. **233**, 702 (1958)
10. Axelrod, J., Vessel, E.S.: Heterogeneity of N- and O-methyl transferases. Molec. Pharmacol. **6**, 78 (1970)
11. Axelrod, J., Weissbach, H.: Enzymatic O-methylation of N-acetylserotonin. Science **131**, 312 (1960)
12. Axelrod, J., Weissbach, H.: Purification and properties of hydroxyindole-O-methyl transferase. J. biol. Chem. **236**, 211 (1961)
13. Baker, P.C., Quay, W.B., Axelrod, J.: Development of hydroxyindole-O-methyltransferase activity in the eye and brain of the amphibian *Xenopus laevis*. Life Sci. **4**, 1981 (1965)
14. Banerjee, S.P., Snyder, S.H.: Methyltetrahydrofolic acid mediates N- and O-methylation of biogenic amines. Science **182**, 74 (1973)
15. Belleau, B., Burba, J.: Occupancy of adrenergic receptors and inhibition of catechol-O-methyl transferase by tropolones. J. med. Chem. **6**, 755 (1963)
16. Bo, P.T., McIsaac, W.M., Tausey, L.W., Kralik, P.M.: Hydroxyindole-O-methyltransferase. II. Inhibitory activities of some N-acetyltryptamines. J. pharm. Sci. **57**, 1998 (1968)
17. Brown, D.D., Tomchick, R., Axelrod, J.: Distribution and properties of a histamine-methylating enzyme. J. biol. Chem. **234**, 2948 (1959)
18. Brownstein, M., Saavedra, J.M., Axelrod, J.: Control of N-acetylserotonin by a beta-adrenergic receptor. Molu. Pharmacol. **9**, 605 (1973)
19. Brownstein, M., Saavedra, J.M., Palkovits, M., Axelrod, J.: Histamine content of hypothalamic nuclei of the rat. Brain Res. In press (1977)

20. Ciaranello, R.D., Axelrod, J.: Genetically controlled alterations in the rate of degradation of phenylethanolamine N-methyltransferase. J. biol. Chem. **248**, 5616 (1973)
21. Ciaranello, R.D., Jacobowitz, D., Axelrod, J.: Effect of dexamethasone on phenylethanolamine N-methyltransferase in chromaffin tissue of the neonatal rat. J. Neurochem. **20**, 799 (1973)
22. Coyle, J.T., Henry, D.: Catecholamines in fetal and newborn rat brain. J. Neurochem. **21**, 61 (1973)
23. Deguchi, T., Barchas, J.: Inhibition of transmethylations of biogenic amines by S-adenosyl-homocysteine, J. biol. Chem. **246**, 3175 (1971)
24. Eisenfeld, A.J., Landsberg, L., Axelrod, J.: Effect of drugs on the accumulation and metabolism of extraneuronal norepinephrine in the rat heart. J. Pharmacol. exp. Ther. **158**, 378 (1967)
25. Enzyme Nomenclature. Amsterdam: Elsevier 1973, pp. 120-126
26. Finkle, B.J., Nelson, R.F.: Enzyme reaction with phenolic compounds by a meta O-methyltransferase in plants. Biochim. biophys. Acta **78**, 747 (1963)
27. Fuller, R.A., Hunt, J.M.: Substrate specificity of phenylethanolamine N-methyl transferase. Biochem. Pharmacol. **4**, 189 (1965)
28. Hoffman, A.R., Ciaranello, R.D., Axelrod, J.: Substrate and inhibitor kinetics of bovine phenylethanolamine N-methyl transferase. Biochem. Pharmacol. In press (1977)
29. Hökfelt, T., Fuxe, K., Goldstein, M., Johansson, O.: Evidence for adrenaline neurons in the rat brain. Acta physiol. scand. **89**, 286 (1973)
30. Inscoe, J.K., Daly, J., Axelrod, J.: Factors affecting the enzymatic formation of O-methylated dihydroxy derivatives. Biochem. Pharmacol. **14**, 1257 (1965)
31. Krakoff, L.R., Axelrod, J.: Inhibition of phenylethanolamine N-methyltransferase. Biochem. Pharmacol. **16**, 1384 (1967)
32. Laduron, P.: N-methylation of dopamine to epinine in brain tissue using N-methyltetrahydrofolic acid as the methyl donor. Nature (New Biol.) **238**, 212 (1972)
33. Landsberg, L., De Champlain, J., Axelrod, J.: Increased biosynthesis of cardiac norepinephrine after hypophysectomy. J. Pharmacol. exp. Ther. **165**, 102 (1969)
34. Marki, F., Axelrod, J., Witkop, B.: Catecholamines and methyl transferases in the South American toad (*Bufo marinus*). Biochim. biophys. Acta **58**, 367 (1962)
35. Palkovits, M., Brownstein, M., Saavedra, J.M., Axelrod, J.: Norepinephrine and dopamine content of hypothalamic nuclei. Brain Res. In press (1977)
36. Quay, W.B.: Retinal and pineal hydroxyindole-O-methyl transferase. Life Sci. **4**, 983 (1965)
37. Saavedra, J.M.: Enzymatic isotopic assay for and presence of beta-phenylethylalanine in brain. J. Neurochem. **22**, 211 (1974)
38. Saavedra, J.M., Axelrod, J.: Psychotomimetic N-methylated tryptamines: formation in brain in vivo and in vitro. Science **175**, 1365 (1972)
39. Saavedra, J.M., Axelrod, J.: A specific and sensitive enzymatic assay for tryptamine in tissues. J. Pharmacol. exp. Ther. **182**, 363 (1972)
40. Saavedra, J.M., Axelrod, J.: Demonstration and distribution of phenylethanolamine in brain and other tissues. Proc. nat. Acad. Sci. (Wash.) **70**, 772 (1973)
41. Saavedra, J.M., Brownstein, M., Axelrod, J.: A specific and sensitive microassay for serotonin in tissues. J. Pharmacol. exp. Ther. **186**, 508 (1973)
42. Saavedra, J.M., Coyle, J.T., Axelrod, J.: Distribution and properties of non-specific N-methyltransferase in brain. J. Neurochem. **20**, 743 (1973)
43. Saavedra, J.M., Palkovits, M., Brownstein, M., Axelrod, J.: Localization of phenylethanolamine N-methyltransferase in the rat brain nuclei. Nature (Lond.) **248**, 695 (1974)
44. Saavedra, J.M., Palkovits, M., Brownstein, M., Axelrod, J.: Serotonin distribution in the nuclei of the rat hypothalamus and preoptic region. Brain Res. In press

30

45. Snyder, S.H., Axelrod, J.: Inhibition of histamine methylation in vivo by drugs. Biochem. Pharmacol. **13**, 536 (1964)
46. Snyder, S.H., Axelrod, J.: Sex differences and hormonal control of histamine methyltransferase activity. Biochim. biophys. Acta **2**, 416 (1965)
47. Taylor, K.M., Snyder, S.H.: Isotopic microassay of histamine histidine, histidine decarboxylase and histamine methyltransferase in brain tissue. J. Neurochem. **19**, 1343 (1972)
48. Thoenen, H., Mueller, R.A., Axelrod, J.: Neuronally dependent induction of adrenal phenylethanolamine N-methyltransferase by 6-hydroxydopamine. Biochem. Pharmacol. **19**, 669 (1970)
49. Wooten, G.F., Saavedra, J.M.: Axonal transport of phenylethanolamine N-methyltransferase in toad sciatic nerve. J. Neurochem. In press (1977)
50. Wurtman, R.J., Axelrod, J.: Control of enzymatic synthesis of adrenaline in the adrenal medulla by adrenal cortical steroids. J. biol. Chem. **241**, 2301 (1966)
51. Wurtman, R.J., Axelrod, J., Fischer, J.E.: Melatonin synthesis in the pineal gland: Effects of light mediated by the sympathetic nervous system. Science **143**, 1328 (1964)
52. Wurtman, R.J., Axelrod, J., Kelly, D.E.: The Pineal. New York: Academic Press, 1968
53. Wurtman, R.J., Axelrod, J., Phillips, L.S.: Melatonin synthesis in the pineal gland: control by light. Science **142**, 1071 (1963)
54. Wurtman, R.J., Axelrod, J., Potter, L.T.: The disposition of catecholamines in the rat uterus and the effect of drugs and hormones. J. Pharmacol. exp. Ther. **144**, 150 (1964)
55. Wurtman, R.J., Axelrod, J., Toch, R.: Demonstration of hydroxyindole-O-methyltransferase, melatonin and serotonin in a metastatic parenchymatous pinealome. Nature (Lond.) **204**, 1323 (1964)
56. Wyatt, R.J., Saavedra, J.M., Axelrod, J.: A dimethyltryptamine-forming enzyme in human blood. Amer. J. Psychiat. **130**, 754 (1973)

On the Development and Utilization of Assays for Biological Transmethylation Involving S-Adenosylmethionine[1]

R.J. BALDESSARINI[2]

For many years, there has been considerable interest in the chemistry and pharmacology of biological transmethylation in the field of neurophsychiatry. This interest was largely stimulated by the fact that many of the natural or synthetic substances which produce hallucinations or other features of psychotic illness are methylated amines – see (5). As early as 1952, Osmond and Smythies reported (73) a suggestion of the biochemist Harley-Mason that abnormal transmethylation of an endogenous amine, possibly dopamine, might produce a psychotomimetic compound like mescaline (3, 4, 5-trimethoxyphenylethylamine). More direct evidence consistent with this hypothesis was the observation that methionine, uniquely among several amino acids, and especially when combined with an inhibitor or monoamine oxidase, led to striking but transient exacerbations of the psychotic symptoms of chronic schizophrenic patients (75). This clinical phenomenon is unique among biochemical findings in schizophrenia in that it has been confirmed by several groups and so far contradicted by none (1, 2, 15, 21, 41, 49, 74, 86; see also 27, 29).

Moreover, a similar result was obtained with betaine, another substance capable of contributing a methyl group to intermediary metabolism in mammalian tissues (22). Also, an unconfirmed report which is not easily interpreted is that methionine-sulfoximine, a metabolic antagonist of methionine, may have had beneficial effects in a small number of schizophrenics (42). Another observation is that schizophrenia-like psychoses appear in unexpectedly high frequencies in patients with homocystinuria and their relatives, and this inborn error is associated with high circulating levels of methionine (25). There have also been reported suggestions that there may be unusual methylated phenylethylamines (37) or indoleamines in the urine of schizophrenic patients (see 36, 70, 71, 89). On the other hand, the significance of the latter findings has been questioned or not supported by several recent studies (30, 43, 57, 72, 82, 93). Nevertheless, reports of abnormal excretion of possibly psychodysleptic N-methylated tryptamines have continued to appear, even with the application of less ambiguous analytical methods (70, 71, 89).

[1] Based in part on material presented previously in: Baldessarini, R.J.: Biological transmethylation involving S-adenosylmethionine: Development of assays and implications for neuropsychiatry. Int. Rev. Neurobiol. **18**, 41 (1945); and Baldessarini, R.J.: Sviluppo ed applicazione dei metodi per la determinazione delle transmetilazioni biologiche S-S-adenosilmetionina dipendenti; in: Transmetilazioni e Sistema Nervoso Centrale, Edizioni Minerva Medica, Roma, 1976, p. 51.

[2] Associate Director, Psychobiological Research Laboratories. The Mailman Research Center, McLean Division of the Massachusetts General Hospital; Associate Professor of Psychiatry, Harvard Medical School, Boston, Massachusetts 02114/USA.

These findings are particularly interesting in the light of the observation that the hallucinogen N, N-dimethyltryptamine may not produce tolerance to its behavioral effects in the cat (38), as it should not if it is an endogenous toxin that contributes to the appearance of chronic psychosis in man. There is also an unconfirmed report of uncertain significance that the ability of blood samples from schizophrenic patients to support the methylation of nicotinamide may be higher than normal (23), as well as an observation of clinical worsening of schizophrenics upon injection of a preparation of a plant extract containing catechol-O-methyltransferase (COMT) activity (39). There have also been suggestions that antipsychotic drugs may inhibit a variety of amine-methylating enzymes (3, 40, 55), but these effects are weak and of dubious functional significance. Another report has suggested that the decarboxylation of labeled dihydroxyphenylalanine (dopa) to dopamine by erythrocytes of schizophrenic patients may be more active than normal (91), thereby possibly increasing the availability of an aromatic amine capable of accepting methyl groups. Interest in the possibility that transmethylation might be abnormal in schizophrenia has also been stimulated recently by reports that the activity of monoamine oxidase (MAO) may be decreased in the blood platelets of schizophrenic patients (69, 94), although apparently not in their brains (80). There is also a recent observation that a methyltransferase dependent on S-adenosylmethionine (SAMe) may be more active in blood platelets of schizophrenic patients than of comparison subjects, possibly due to the decreased availability of a dialyzable inhibitor of the enzyme in schizophrenics (95). This enzyme appears to be similar to the nonspecific N-methyltransferase which occurs in many tissues along with a dialyzable inhibitor (78); it has even been reported to occur in low activity in human brain tissue (61), although it is probably not increased in activity in the brains of schizophrenics (32). In the affective disorders there is also considerable, though somewhat inconsistent, evidence to suggest that there may be abnormality of amine metabolism (see 7), including an unconfirmed report of decreased activity of erythrocyte COMT in depressed women (28). There is also a preliminary report of abnormal metabolism of methionine in schizophrenic and depressive states, as estimated by the rate of appearance of radioactive CO_2 in the breath following intravenous injection of [^{14}C-methyl]-labeled methionine (48). Recently there have been preliminary studies suggesting that injections of SAMe may be of therapeutic benefit to depressed patients, by an uncertain mechanism (35). The weight of these various observations has supported the idea that studies of transmethylation of biogenic amines in the major mental illnesses might be of some importance in the attempt to understand their pathophysiology, and possibly lead to insights into their causes and more effective treatment.

The observations relating to the unique exacerbation of psychosis when patients were treated with methionine (or betaine), with or without an inhibitor of MAO, but not with other amino acids, strongly suggested that methionine might be acting as a methyl donor. This would happen after its conversion, with ATP, by methionine adenosyltransferase, to the important methyl donor S-adenosylmethionine (SAMe). Some aspects of this topic have been approached by studies of the physiologic chemistry of SAMe. An initial problem was the need for a sensitive and specific assay for tissue levels of this methyl donor. One approach to this problem resulted in the development of a double-isotopic enzymatic assay for SAMe (13, 14, 50). The basic principle involved is the isotope dilution of radioactive SAMe with the endogenous compound present in acid extracts of a tissue, and estimation of the specific radioactivity of the diluted SAMe by the enzymatic formation of melatonin from the methyl donor and N-acetylserotonin. The specificity of the assay depends on the selectivity of the enzyme hydroxyindole-O-methyltransferase (HIOMT) for SAMe as methyl donor and the absence of appreciable amounts of N-acetylserotonin in most tissues (with the notable exception of the pineal gland). The assay could be conducted with methyl-labeled SAMe as the only labeled cosubstrate, but preliminary experiments revealed that the efficiency of production of melatonin was low and somewhat variable, and failed to yield a linear relationship between the amount of unlabeled SAMe present and the amount of melatonin produced. Thus, in order to monitor the efficiency of the production of melatonin, a second label was introduced in the acetyl group of the methyl-accepting cosubstrate, N-acetylserotonin. Ordinarily, [^{3}H-acetyl]-N-acetylserotonin and [^{14}C-methyl]-SAMe are used, largely so as to take advantage of the relative chemical stability of the [^{14}C]-labeled SAMe. However, when assays of relatively low concentrations of SAMe are required, as in blood specimens, it is advantageous to increase the sensitivity by reversing the labels and to use tritiated SAMe of high specific radioactivity and [^{14}C]-labeled N-acetylserotonin (62). It can be predicted mathematically that the ratio of the two labels in the recovered melatonin should be linearly related to the amount of unlabeled SAMe present, and this prediction has been verified experimentally (4). More recently, the principle of this technique has been applied in a chromatographic assay for SAMe, which is elegant in its simplicity (79). In the chromatographic assay, again radioactive SAMe is added to acid homogenates of tissue to establish the specific radioactivity, and SAMe is recovered by Dowex-Na$^+$ ion-exchange chromatography; the specific activity of SAMe in sulfuric acid eluates as estimated by counting and by spectrophotometric assay of adenine compounds, is proportional to the level of endogenous SAMe. Estimates of tissue levels of SAMe by this method agree quite well with those provided by the enzymatic method, although they are generally somewhat lower (as much as 50%), possibly owing to greater purity of the authentic SAMe used to establish standard curves for the assays.

The materials required for the enzymatic assay of SAMe include partially purified methyltransferase enzyme (HIOMT) prepared from beef pineal gland, which is readily available from commercial sources. The methyl acceptor, N-acetylserotonin, is easily and quickly prepared by reacting serotonin with radioactive acetic anhydride in a mildly alkaline medium, and separating the products by preparative paper chromatography. SAMe, either unlabeled or radioactively labeled with ^{14}C or ^{3}H, is also readily available commercially. The tissue is extracted with trichloroacetic acid, and the labeled SAMe can be introduced directly into the homogenates to avoid problems of recovery or losses of the endogenous SAMe by establishing the specific radioactivity of the SAMe immediately. Even without this precaution, the recovery of authentic SAMe is virtually quantitative (>95%). The samples can then be frozen and assayed later at one's convenience. Large numbers of samples can be handled easily at one time. The materials can be prepared in advance and kept frozen, and are stable for many months. When the SAMe preparations, methyl acceptor, and HIOMT are reacted, the product, doubly labeled melatonin, is recovered by extraction into chloroform; the organic phase is washed with NaOH solution, and then counted for ^{3}H and ^{14}C. Quantitative recovery of the product is not required since the assay depends merely on the ratio of the two labels, and it is important only to recover sufficient melatonin for counting and to be certain that melatonin is the only labeled molecule recovered. The authenticity of the recovered product was verified by chromatography in several solvent systems with authentic melatonin. Furthermore, it was shown that negligible radioactivity was recovered by incubation of the methyl acceptor and methyl donor with tissue extracts in the absence of HIOMT, or incubation of labeled SAMe with tissue extracts and HIOMT. Thus, the tissue extracts do not have any significant amounts of HIOMT activity or of substances which accept methyl groups and are extracted into chloroform under the conditions of the assay; moreover, the contribution of methyl acceptors by the partially purified and dialyzed HIOMT preparation is also insignificant. Of several potential methyl donors, only SAMe was found to yield melatonin under the conditions of the assay, although it appeared that S-adenosyl*ethionine* (not normally present in tissue, but found after treatment with high doses of ethionine) can transfer its ethyl group to N-acetylserotonin in the presence of HIOMT. The radioenzymatic assay method is capable of detecting as little as 0.5 n mol of SAMe, when [14C]-SAMe is used, and the use of [3H]-SAMe increases the sensitivity by about an order of magnitude. The precision of the assay is very high.

Measurable quantities of endogenous SAMe were detected in all tissues examined (Table 1), with highest levels found in the adrenal and pineal glands. Most tissues contained from 10 to 50 μg/g of wet tissue, while blood or serum contained 0.5 to 1.0 μg/ml. Brain tissue contained about 10 to 15 μg/g, with no impressive regional distribution. The concentrations of SAMe in brain and liver tended to fall as a function of age in rats. These values may all be slightly high since commercially available authentic SAMe was used as a standard without further repurification and it is now known to be <90% pure.

It is also possible to utilize this assay to estimate the turnover of SAMe in tissues in vivo by introducing labeled L-methionine by intravenous administration and allowing

Table 1. Tissue concentrations of S-adenosylmethionine

Tissue	μg/g or ml	μM[a]
Adrenal	48	200
Pineal	38[b]	158
Liver	29	121
Heart	26	108
Spleen	24	100
Kidney	20	83.3
Lung	11	45.8
Brain	11	45.8
Whole blood (human)	1.0	3.3
Leukocytes (human)	1.5	6.3
Lymphocytes (human)	2.2	9.1
Serum	0.5	1.2

Data are mean values for n = 3 to 20 assays with rat tissues, except as noted otherwise, and are summarized from Refs. 9, 13, 14, 62.

[a] An approximate value based on assumption that SAMe distributes in tissue water (60% of net wt.).

[b] An approximate value not corrected for the effect of endogenous N-acetylserotonin.

the tissues to generate labeled SAMe. The assay can then be used to estimate the specific activity of SAMe over time, and this activity will change as a function of the rate of synthesis and utilization of endogenous SAMe (14). In this way, it was found that the rate of production of labeled SAMe in rat liver was extremely rapid: labeled SAMe was detected in less than 5 min and its half-life was of the order of 10 min; brain was found to produce SAMe somewhat less rapidly and to consume it much more slowly than liver, and this difference correlates with the much lower activity of methionine adenosyltransferase in brain (14, 63).

Up to this point, the most important conclusion to be drawn from the preliminary survey of tissue concentrations of SAMe was that this important metabolic intermediary accumulates to easily detectable levels, and that it is neither so labile nor so rapidly utilized as to be unmeasurable. Moreover, the rate of synthesis and turnover of SAMe occurs with a time course which can easily be measured in vivo. The next aspects of the metabolism of SAMe to be considered were whether its availability might be enhanced by increased input of methionine, or, conversely, whether its utilization might be increased by increased availability of methyl acceptors. It has been demonstrated in the rat in vivo that intravenously injected [^{14}C]-methyl-labeled methionine results in the formation of methyl-labeled catecholamine metabolites such as vanillylmandelic acid (VMA) (3). It was found that the systemic administration of methionine in the rat produced striking increases of SAMe in liver and other peripheral tissues, and smaller increases in the brain as well (6) (Table 2A) and this finding has been confirmed independently in the rabbit (79) and the rat (77). Increased availability of the precursor methionine probably can enhance the rate of production of SAMe, since the normal tissue levels of methionine (50-90 nmol/g, or about 80-150 μM) (59, 77) are close to the K_m for methionine (about 90 μM) (63) for methionine adenosyltransferase, which is thus probably not normally saturated with substrate. These observations were thus consistent with the hypothesis that the administration of methionine to psychiatric patients might increase the availability of SAMe for amine methyltransferase reactions. It was also noted when end-to-side portacaval venous anastomoses were made surgically in the rat, that there was a marked decrease in the concentration and total amount of SAMe in the liver (10). This effect was apparently not due to nonspecific toxic effects on the liver, since the activity of methionine adenosyltransferase was not deficient in the "shunted" livers, and furthermore, when more of the substrate methionine was made available by systemic administration, the shunted rats were able to produce nearly as much hepatic SAMe as their "sham-operated" controls (Table 2B). Thus, it appeared likely that the production of SAMe in liver was highly dependent on the availability of the precursor L-methionine.

Other experiments have also provided more indirect evidence which is consistent with the same conclusions. For example, it was found that human leukocytes from patients with chronic myelocytic leukemia have striking elevations of SAMe levels (9), while the activity of methionine adenosyltransferase was not different from that of normal white blood cells (8). It was later reported that similar leukemic white blood cells had greater than normal uptake of labeled methionine in vitro (84). These findings taken together suggest that the increased SAMe levels in leukemic cells might be due to increased availability of the precursor, methionine. More recently, it was noted that SAMe levels in rat brain (26, 92) (Table 3) and in human blood (64) are depleted by large doses of L-dopa, which is rapidly methylated by catechol-O-methyltransferase (COMT) with SAMe. One of the findings

Table 2. Tissue content of S-adenosylmethionine after treatment with methionine (mean content ± SEM)

A. Tissues of intact rats (SAMe, μg/g)

Tissue	Control	Methionine-treated (100 mg/kg, i.p.)
Liver	26.0 ± 1.0	115.0 ± 15.0
Heart	25.7 ± 6.2	42.8 ± 0.6
Brain	11.5 ± 0.5	15.0 ± 0.6

B. Liver at 6-8 weeks after end-to-side portacaval venous anastomosis (SAMe, μg/liver)[a]

Condition	Control	Methionine-treated (200 mg/kg, i.p.)
Sham-operated	330 ± 15	875 ± 75
Portacaval shunted	111 ± 14	775 ± 75

[a] Concentrations are expressed per organ since liver weight was slightly decreased after shunting. (Refs.: 5, 6, 10).

Table 3. Effects of methyl acceptors on levels of S-adenosylmethionine

Substance	Dose (mg(kg, i.p.)	SAMe (% of control ± SEM)	
		Liver	Brain
L-dopa (acute)	10	——	95.2 ± 6.5
	30	——	63.6 ± 1.8*
	100	108.0 ± 9.5	32.7 ± 2.4*
L-dopa (chronic)	100	78.8 ± 6.1*	16.7 ± 2.0*
Pyrogallol	100	30.4 ± 1.8*	25.0 ± 5.0*
Purpurogallin	100	35.3 ± 8.0*	70.7 ± 4.0*
Nicotinamide	100	111.0 ± 1.1	——

*$P < 0.05$ or less for $N > 6$ rats

Drugs were given 30-60 min before sacrifice after an acute dose or the last dose of repeated treatment for 10 days. (Refs.: 6, 26, 92).

in these experiments was that several hours after a dose of L-dopa, the reduced levels of SAMe in the rat brain had not only returned to normal, but were actually 30-40% higher than normal (26). A possible explanation for this change follows from the observation that 3-O-methyl-dopa (3-methoxytryosine) accumulates in brain and disappears with a time course which is strikingly similar to the observed increase of SAMe (18). Moreover, when isolated nerve endings were preloaded with this methylated derivative of L-dopa, the uptake of labeled methionine was markedly enhanced, suggesting that methionine can be taken up in exchange for other amino acids (12). This phenomenon might provide for increased availability of methionine to cells containing adenosyltransferase, and might at least partly explain the increased levels of SAMe several hours after an acute dose of L-dopa.

In addition to the SAMe-depleting effects of L-dopa, it had been noted earlier that a variety of polyphenolic substrates for COMT can also lead to the depletion of tissue levels of SAMe (6) (Table 3). These compounds evidently are able to decrease tissue levels of SAMe by increasing its utilization beyond the capacity of synthesis to keep pace with the demand for new molecules of the methyl donor. The effect of L-dopa and of polyphenols was more striking in the brain, a tissue with relatively less ability to produce SAMe, than in the liver, although even in liver large doses of these agents or their repeated administration did lead to the partial depletion of hepatic SAMe (Table 3). These results were not associated with decreased activity of methionine adenosyltransferase in the case of one polyphenolic substance, pyrogallol (6). Thus, these findings indicate that the availability of SAMe is dependent not only on the availability of methionine, but also on the demands of methyl acceptors for methyl groups. It also follows that the rate of production of methylated metabolites might increase after large doses of methionine, as required by the hypothesis that methionine produces exacerbations of psychosis in schizophrenic patients by increasing the availability of certain methylated products, possibly including psychotogenic amines. An interesting negative observation among these experiments was that even a large dose (100 mg/kg, i.p.) of nicotinamide failed to alter hepatic concentrations of SAMe (Table 3), and there was no effect on liver or brain SAMe even when nicotinamide was given intravenously (6). This lack of an SAMe-depleting action of nicotinamide, which is known to be methylated by SAMe (24), is of some interest in the light of the highly controversial proposal that nicotinamide or nicotinic acid might be of value in the So-called "orthomolecular" treatment of schizophrenia, and that such an effect might be mediated by the action of nicotinamide as an acceptor of methyl groups (44, see also 58). There is even evidence that methionine might decrease the excretion of N-methylnicotinamide (45, 87). It is also interesting that large doses of nicotinic acid failed to prevent the exacerbation of psychosis following methionine and an MAO inhibitor (15, see also 58).

The possible clinical-metabolic significance of the administration of methionine or methyl acceptors is still not well known. However, the currently available evidence concerning the possibility that methionine may increase the methylation of biogenic amines in patients is generally not supportive of the hypothesis that methionine produces exacerbation of psychosis in schizophrenia by increasing the availability of psychotogenic methylated amines. Thus, in one report, while the urinary excretion of metanephrine and normetanephrine was markedly increased by treatment with an MAO inhibitor, and apparently increased even somewhat more on the addition of methionine (20 g), the effect of methionine was not quite significant statistically (49). Also, large doses of methionine plus tryptophan failed to increase the excretion of vanillylmandelic acid (VMA) (20). In another study, when large doses (up to 20 g) of methionine without an inhibitor of MAO were given to schizophrenic patients, there was a clear exacerbation of psychosis, but no increased excretion of the methylated catecholamine metabolites, VMA and methoxyhydroxyphenylethyleneglycol (MHPG) (2). A similar failure to find increased excretion of a number of O- and N-methylated metabolites after 10 g of methionine has been reported more recently (29). There have been suggestions that certain methylated tryptamines might be excreted in abnormally high quantities in schizophrenic patients given large doses of methionine (86); however, these findings were obtained with analytical methods of very low specificity, and furthermore, they are inconsistent with other results (88). One of the problems with such studies is that methionine as well as other amino acids, including the precursors of methionine, cysteine and homocysteine, appear to increase the excretion of indoleamines generally, particularly tryptamine, evidently by altering the metabolism of tryptophan (87). This effect of cysteine has been used to increase the excretion of tryptamines in schizophrenic patients, and the administration of cysteine plus an inhibitor of MAO has been reported to increase the excretion of N-methylated tryptamines in such patients (89), but again, the significance of these findings is not clear and they were obtained with assay methods which are not unambiguous. To date, the effects of methionine-loading in schizophrenic and other subjects upon the excretion or blood levels of N-methylated indoleamines, as measured by recently developed very specific and sensitive gas chromatographic and mass spectrometric methods (57, 70, 71, 93), has not been reported.

There are several other findings which are generally not supportive of the idea that methylation of amines may be abnormal in schizophrenia or other psychoses. For example, mean circulating levels or activities of SAMe, methionine adenosyltransferase, and of COMT have all been reported to be statistically within the range of values obtained with normal or other comparison groups (34, 62). Nevertheless, the existence of a subgroup of schizophrenics not obviously identified on clinical grounds alone, but having an unusual ability to methylate amines remains a possibility (62). When a large number of subjects were challenged with the methyl acceptor, protocatechuic acid, there was no difference in the excretion of methylated products by the schizophrenics (76), although such tests should be repeated with precursors of the indoleamines and the newer assays for their methylated products, and they should also be repeated with and without methionine loads. In another loading-type experiment, nonschizophrenic subjects given huge doses of amphetamine failed to excrete the hallucinogenic substance p-methoxy-amphetamine, although the authentic exogenous substance could be detected by the methods employed (81); on the other hand, this finding is not directly relevant to the idiopathic psychoses, and even the psychosis which follows large doses of amphetamine need not be the result of a psychotomimetic metabolite. There has been some hope that substances which accept methyl groups would have clinically beneficial effects in schizophrenia. Unfortunately, most of the materials available are toxic and unsuitable for clinical use. L-dopa has been given to schizophrenic patients, but its central excitatory effects are evidently so great as to exacerbate the psychosis (68, 97). Polyphenolic compounds inhibit catechol-O-methyltransferase, and can accept methyl groups from S-adenosylmethionine (11). While most of these substances are toxic, butylgallate has been given in doses up to 5 g to chronic schizophrenics but with disappointing results, including gradual clinical worsening, possibly due to coincidental deprivation of antipsychotic medications, and evidence of hepatic toxicity at doses over 2 g a day (83). There was no evidence that the drug was in fact an effective inhibitor of methylation in the patients; even if this did occur, it is not likely that doses below the toxic level of 2 g a day have an important SAMe-depleting effect, judging by analogy to the relatively mild and short-lasting effects of large doses (several grams a day) of L-dopa on blood concentrations of SAMe in Parkinson patients (64).

Another biochemical consideration also casts some doubt on the hypothesis that methionine loads might actually increase the rate of methylation of amines. The data in Table 1 indicate that the concentration of SAMe, at least in rat tissues, ranges from about 50 to 200 μM. Since the half-maximal velocity of most of the known amine-methyltransferases occurs at SAMe concentrations (K_m) of about 5 to 50 μM (19, 31, 53, 78), it seems probable that additional SAMe will not force the reactions to an important extent. Thus, it would be well to consider alternative metabolic effects of high doses of methionine by which to explain its effects on schizophrenia. One interesting possibility is suggested by the recent observations that S-adenosylhomocysteine, an end product of methyltransferase reactions utilizing SAMe, exerts a strong competitive inhibitory action on several amine methyltransferase reactions in μM concentrations (19, 31, 54, 98). Thus, it is conceivable that increased levels of SAMe produced by large doses of methionine might alter the ratio SAMe: S-adenosylhomocysteine to favor the methylation of amines.

A provocative aspect of transmethylation reactions which has recently been widely discussed is that certain amine methyltransferase reactions might be supported preferentially by N^5-methyltetrahydrofolate (MTHF) rather than SAMe, including the direct N-methylation of dopamine (followed by β-hydroxylation) as a possible alternative to the synthesis of epinephrine by way of norepinephrine (52). More recently, MTHF was reported to be a possible and even a preferred cosubstrate for the transfer in vitro of methyl groups to nitrogen or oxygen atoms of certain indoleamines, including tryptamine (16, 46). The possible physiologic significance of these proposed reactions is not clear. A biochemical consideration that tends to cast some doubt on their importance is that the K_m for the MTHF in such reactions is similar to the values for SAMe in many methyltransferase reactions (about 10-25 μM) (16, 53), while the concentrations of the methylfolate in mammalian brain are much lower (never more than about 1 μM, and in most regions several orders of magnitude less); nevertheless, the highest levels have been reported, interestingly, to occur in regions rich in endogenous indoleamines (51). These findings suggest either that the tissue concentrations of MTHF are likely to be an important factor limiting the rate of production of methylated indoleamines, or that these reactions are not very important in vivo. It is not yet clear whether large doses of substances, including methionine and betaine, capable of donating methyl groups to the "one-carbon" metabolic pool from which MTHF acquires its methyl group, can increase the availability of MTHF in the mammalian brain, and this possibility should be investigated. Recent information has tended to cast considerable doubt on the importance of MTHF in the transfer of methyl groups to oxygen or nitrogen atoms of indoleamines or catecholamines. Thus, it has been noted that some of the products of reaction of MTHF with indoleamines are not N-ω-methylated products (55). Furthermore, it now appears likely that cyclic derivatives incorporating the amino group can form from indoleamines (tetrahydro-β-carbolines) in the presence of MTHF (17, 47, 53, 60, 66, 67, 96). These cyclic derivatives may represent products of nonenzymatic condensation with formaldehyde (47, 67) which can be produced enzymatically in some tissues from MTHF (33), and in some tissues also from SAMe (66). It is also suspected (55) that the proposed N-methylation of dopamine (53) and the O-methylation of 5-hydroxyindoles by MTHF

(16) may also have been confused with condensation reactions in the presence of formaldehyde formed from MTHF. Furthermore, the question whether formaldehyde may form from SAMe in tissues other than erythrocytes (66) is not resolved. Presumably, the formation of formaldehyde from SAMe is related to the ability of erythrocyte and pituitary enzymatic activity to form methanol from SAMe (4).

In addition to the assay of SAMe by the radioactive-enzymatic method, analogous concepts and techniques also permit the measurement of other important metabolic components of methylating systems (Table 4). For example, if the specific radioactivity of SAMe in the reaction described above for the assay of SAMe is held constant, and that of the labeled cosubstrate, N-acetylserotonin is decreased by the addition of the unlabeled authentic indole, there again results a linear relationship between the radioactivity ratio in the product of methyl transfer, melatonin, and the amount of N-acetylserotonin present (Baldessarini, 1965, unpublished observations). In principle, this strategy might be used for the assay of tissue levels of a variety of methyl acceptors in addition to the precursor of melatonin. The requirements of such assays would include sufficient activity of a methyltransferase enzyme so as to yield detectable quantities of a methylated product, and sufficient selectivity of the partially purified enzyme preparation so that it methylates only the substrate of interest, and not other methyl acceptors in the tissue. Alternatively, relatively crude enzyme preparations may suffice if simple and efficient separation techniques will permit the isolation of only a single methylated product. Even the presence of large amounts of an alternative methyl acceptor (e.g., after treatment with large doses of L-dopa) should not interfere with such assays, unless the competing reaction effectively prevented the recovery of measurable radioactivity in the product desired; even gross alterations in the kinetic efficiency of the assay reaction produced by tissue components or drugs should not matter since only a ration of ^{3}H: ^{14}C in the final methylated product, and not its quantitative recovery, is required. Such an assay has been developed for tissue histamine (85). This method is highly sensitive and specific and it a voids a problem that has beset previous bioassay and spectrophotofluorimetric assays of histamine: the occurrence of spuriously high values due to the presence of other polyamines, notably spermidine, and particularly in assays of blood and brain tissue. In this assay, SAMe is present at constant specific radioactivity (labeled in the methyl group with ^{14}C or ^{3}H), and is reacted with labeled histamine (with the alternative radioisotope), buffered extracts of tissue (which have been boiled to inactivate enzymes and endogenous SAMe) containing endogenous histamine (which survives boiling), and partially purified histamine N-methyltransferase prepared from guinea pig cerebral cortex. The doubly labeled radioactive product, N-methylhistamine, is then extracted into chloroform from the NaOH-treated reaction mixtures as the only labeled product recovered, and the ratio of the two labels is determined. This ratio was found to bear a theoretically predicted linear relationship with the amount of unlabeled histamine present. More recently, several technical modifications have been made to increase the sensitivity of the enzymatic assay for histamine, and it is now possible routinely to measure about 50 pg of tissue histamine, and under some conditions, perhaps even less (56, 90). For many purposes, it is not even necessary to utilize two labels, and one can take advantage of the high specific radioactivity of [^{3}H]-SAMe for a very sensitive single-label microassay for histamine (90). These methods are sufficiently specific and sensitive to permit the assay of histamine in small regions of the brain, including the human brain post-mortem, in which high concentrations of histamine are found in the hypothalamus (about 1 μg/g), the colliculi, certain midbrain nuclei, and the olfactory tubercles (0.1-0.3 μg/g) (56). Blood histamine can also be measured in small samples in this way and these methods are being applied to clinical studies (Lipinski, in preparation).

Variations on the same chemical theme can also be utilized to assay the enzyme which synthesizes SAMe, methionine adenosyltransferase (ATP: L-methionine-S-adenosyltransferase, EC 2.4.2.13) (63). Alternative methods for the assay of this enzyme are based on the principle that labeled methionine and the product SAMe have dissimilar binding to ion-exchange resins or papers, thus permitting chromatographic separation of substrate from product, and an attempt must be made to recover the product quantitatively (see 63, 65). In the first step of the enzymatic double-isotopic assay, [^{3}H-methyl]-L-methionine is reacted with an excess of ATP, Mg^{++}, reduced glutathione, and various amounts of tissue preparations containing methionine adenosyltransferase activity; in the middle of this reaction, [^{14}C]-SAMe is introduced to provide an automatic means of cerrecting for losses of newly synthesized SAMe which might take place even as it is being synthesized. Next, the reaction mixture is deproteinized by the addition of trichloroacetic acid, which is removed by washing with ether, and the second step of the assay is conducted with unlabeled N-acetylserotonin and partially purified beef pineal hydroxyindole-O-methyltransferase (HIOMT) as in the assay for SAMe described above (Table 4). In this case, the ratio ^{3}H: ^{14}C in the doubly labeled melatonin finally recovered in a chloroform extract was predicted and found to be linearly related to the amount of SAMe synthesized in step one, and hence to the activity of the adenosyltransferase (63). The advantage of this method is that it provides for correction for losses of the rather unstable product, SAMe, not only after the reaction is terminated, but also during the assay itself, when reactions leading to the utilization and destruction of SAMe can occur, although they are almost never taken into account in the traditional chromatographic techniques for separating and recovering SAMe.

Finally, in addition to the assay of SAMe and of the enzyme which synthesizes it, as well as of molecules which accept methyl groups from SAMe, it has recently been demonstrated that similar principles can be applied to the assay of L-methionine by an enzymatic derivative, double isotope method (59) (Table 4). In this assay, [^{3}H-methyl]-L-methionine is reacted with [8-^{14}C]-ATP, Mg^{++}, glutathione, and an excess of bacterial methionine adenosyltransferase purified from E. coli; tissue extracts containing endogenous methionine are also included after their treatment on small columns of Dowex anion-exchange resin to remove endogenous ATP. After the enzymatic reaction is terminated, newly synthesized, doubly labeled SAMe is separated from the radioactive cosubstrates on columns of Dowex NH_4^+ cation-exchange resin which are washed with large volumes of water; finally, the doubly labeled SAMe is eluted with NH_4OH and the radioactivity ratio ^{3}H: ^{14}C is counted. It was found to bear the predicted linear relationship to the amount of authentic L-methionine present in the reaction mixture. The method is routinely capable of detecting 150 ng of methionine easily, and can be made sensitive enough to detect 1.5 ng. The results of this assay were found to accord well with the data provided by a Beckman amino acid analyzer, demonstrating in rat tissue, for example, levels of 80-86 nmol/g in liver and 65-68 nmol/g in brain (59).

In summary, the principle of using two radioactive substrates with dissimilar labels has permitted the development of highly sensitive and specific enzymatic assays for estimating tissue levels or activity of several important components in methyltransferase systems in mammalian tissues: L-methionine, S-adenosylmethionine, L-methionine adenosyltransferase, and molecules which accept methyl groups from SAMe, including histamine (Table 4). These methods provide linear standard plots of radioactivity ratios (^{3}H: ^{14}C) vs. the amount of authentic material being assayed. While they involve the utilization of relatively exotic materials, these are either commercially available or easiliy prepared and are stable for many month of use. The assays themselves are relatively simple and rapid and can be done in large numbers. One advantage of several of them is that corrections for recovery and reaction efficiency are provided automatically. The methods have been successfully applied to various fractions of human blood and so they are appropriate for use in clinical metabolic experiments.

There are also several clinically important points which deserve further emphasis. The observation that 10 to 20 g loading doses of methionine, but not several other amino acids, uniquely led to exacerbation of psychosis in schizophrenic patients is highly unusual among biological findings in severe psychiatric illness in that it has been independently confirmed repeatedly in the past decade. The initial suggestion that methionine might act by increasing the production of potentially psychotomimetic methylated amines by way of increased availability of S-adenosylmethionine, is only *partially* sup-

Table 4. Double-label enzymatic assays related to methylation

Assay	Reaction	References
L-methionine (Me)	Me + [^{3}H]–Me + [^{14}C]–ATP + Mg^{++} $\xrightarrow{\text{MAT}}$ [^{3}H, ^{14}C]–SAMe + PP$_i$ + P$_i$	(59)
S-adenosylmethionine (SAMe)	SAMe + [^{14}C]–SAMe + [^{3}H]–NAS $\xrightarrow{\text{HIOMT}}$ [^{14}C, ^{3}H]–MEL	(14)
Methionine adenosyl transferase (MAT)	1.) [^{3}H]–Me + ATP + Mg^{++} $\xrightarrow{\text{Tissue MAT}}$ [^{3}H]–SAMe + PP$_i$ + P$_i$ 2.) [^{3}H]–SAMe , [^{14}C]–SAMe } + NAS $\xrightarrow{\text{HIOMT}}$ [^{3}H, ^{14}C]–MEL	(63)
N-Acetylserotonin (NAS)	[^{14}C]–SAMe + [^{3}H]–NAS + NAS $\xrightarrow{\text{HIOMT}}$ [^{14}C, ^{3}H]–MEL	(Baldessarini, unpublished observations)
Histamine (HIST)	HIST + [^{3}H]–HIST + [^{14}C]–SAMe $\xrightarrow{\text{HNMT}}$ [^{3}H, ^{14}C]–Me–HIST	(85)

The abbreviations used are: Ac: acetyl; HIST: histamine; HIOMT: hydroxyindole-O-methyltransferase; HNMT: histamine-N-methyltransferase; MAT: ATP-L-methionine adenosyltransferase; Me: L-methionine; MEL: melatonin; NAS: N-acetylserotonin; SAMe: S-adenosyl-L-methionine.

ported by the available evidence. Thus, there is excellent metabolic evidence in animals, but very little in man, that the availability of SAMe is dependent on the availability of its precursor, methionine, and on the rate of its utilization, which can be increased by giving large doses of certain methyl acceptors. On the other hand, the evidence that methionine loading increases blood or tissue levels of SAMe in patients remains untested, even though the means of doing so have been available for ten years. Furthermore, the attempts to document increases of the methylation of amines after methionine loading in animals or man have variously been unsuccessful (catecholamines), or when successful (indoleamines) have not been studied with unambiguous analytical methos, or have not been confirmed by independent replication. The more recent introduction of SAMe for parenteral administration in clinical trials suggests yet another more direct approach to the question of whether increased availability of this methyl donor can increase the methylation of amines. The more recent clinical experiments with blood platelets which suggest that there may be either increased synthesis of methylated amines by an indoleamine N-methyltransferase or decreased catabolism of amines by MAO are important findings urgently in need of independent corroboration.

One of the common experimental strategies which surely must help to obscure metabolic idiosyncracies of certain psychotic patients is to study groups of schizophrenic or depressed or manic patients as if "schizophrenia" or "manic-depressive illness" were unitary phenomena. Even the subclassification of patients on the basis of rigorous clinical criteria, as has become more popular for the affective disorders recently, may not be sufficient, particularly among the schizophrenic disorders. If the clinical groups studied are in fact highly heterogeneous biologically, it is then almost inevitable that specific biologic measures of mean differences between groups will fail to detect even significant biological differences unless these are enormous and obvious. One alternative strategy is to take advantage of a given biological "fact" as a starting point from which to acquire further relevant biological information. A striking but neglected example of such an opportunity is the observation that methionine loads induce transient psychotic exacerbations in only a limited proportion of schizophrenic patients, perhaps 50-75% of chronic patients so tested. It is reasonable to study such patients much more intensively, and to compare them not only with normal subjects or patients in other diagnostic categories, but also with other schizophrenics who are clinically similar except that they did not become more psychotic with large doses of methionine. Progress in this direction has been slow in recent years, perhaps to a considerable extent owing to the recent tendency towart excessive conservatism in regard to human experimentation in many countries, and particularly in psychiatry. Clearly, there are very compelling ethical dilemmas presented by the type of experimentation being suggested: one does not consider an experiment which deliberately seeks to provoke even a transient exacerbation of illness casually, particularly when it requires the gaining of "informed consent" from patients whose mental capacity is reduced reason of chronic psychosis. Nevertheless, to conclude that such experimentation should not be done under any circumstances seems short-sighted and irrational. Once such patients as the "methionine responders" are identified among a population of schizophrenics, they should not only be studied metabolically, but they should also be given experimental treatments specifically aimed at decreasing methylation or the formation of certain aromatic amines, as they become available. At the present time there are several drugs which have already been developed at least to the stage

of clinical trials, including inhibitors which selectively block the synthesis of catecholamines (α-methyltyrosine) or 5-hydroxyindoleamine (p-chlorophenylalanine), inhibitors
of aromatic amino acid decarboxylases (MK-486, RO-44602), inhibitors of phenylethyl-
amine-β-hydroxylase (disulfiram, fusaric acid), and an acceptor of methyl groups from
SAMe which inhibits at least COMT (butylgallate). It might also be possible to devise
other relatively nontoxic strategies to interfere with methylation reactions; for example,
guanidinoacetic acid is a natural metabolite which utilizes a large number of methyl
groups daily in the synthesis of the biologically almost inert product creatine, and it
might be given safely in large doses (Matthysse, personal communication). In attempting
to grapple with the ethical issues raised by these experiments, the risk of short-term exacerbation of an already chronic and virtually incurable illness such as dementia praecox
should be balanced against the chance of offering more rational, though experimental,
treatments as well as the considerable nonspecific benefits that may derive from the in-
tensive study and treatment of patients who might otherwise receive little more than
custodial care. In short, the effect of methionine in schizophrenia and the several metabolic findings related to the synthesis and degradation of methylated indoleamines in
that syndrome are among the very few promising clues to a pathophysiologic substratum
of a major psychiatric illness that are based on *clinical* metabolic experimentation, and
as such, they can and should be pursued vigorously.

References

1. Alexander, F., Curtis, G.C., Sprince, H., Crosley, A.P.: L-methionine and L-trypto-
   phan feedings in non-psychotic and schizophrenic patients with and without tranylcypromine. J. nerv. Dis. **137**, 135 (1963)
2. Antun, F.T., Burnett, G.B., Cooper, A.J., Daly, R.J., Smythies, J.R., Zealley, A.K.:
   Effects of L-methionine (without MAOI) in schizophrenia. J. psychiat. Res. **8**, 63
   (1971)
3. Antun, F., Eccleston, D., Smythies, J.R.: Transmethylation process in schizophrenia.
   In: Brain Chemistry and Mental Disease, Ho, B.T., McIsaac, W.M. (eds.) New York:
   Plenum, pp. 61-71
4. Axelrod, J., Cohn, C.K.: Methyltransferase enzymes in red blood cells. J. Pharmacol.
   exp. Ther. **176**, 650 (1971)
5. Baldessarini, R.J.: Factors influencing S-adenosylmethionine levels in mammalian
   tissues. In: Amines and Schizophrenia, Himwich, H., Kety, S., Smythies, J. (eds.)
   Oxford: Pergamon 1966, p. 199
6. Baldessarini, R.J.: Alterations in tissue levels of S-adenosylmethionine. Biochem.
   Pharmacol. **15**, 741 (1966)
7. Baldessarini, R.J.: Biogenic amine hypotheses in affective disorders. In: The Nature
   and Treatment of Depression, Draghi, S.C., Flach, F.F. (eds.) New York: Wiley,
   1975. Chap. 9, p. 347
8. Baldessarini, R.J., Bell, W.R.: Methionine-activating enzyme and catechol-O-methyltransferase activity in normal and leukemic white blood cells. Nature (Lond.) **209**,
   78 (1966)
9. Baldessarini, R.J., Carbone, P.P.: Adenosylmethionine elevation in leukemic white
   blood cells. Science **149**, 644 (1965)
10. Baldessarini, R.J., Fischer, J.E.: S-adenosylmethionine following portacaval anasto-
    moses. Surgery **62**, 311 (1967)

11. Baldessarini, R.J., Greiner, E.: Inhibition of catechol-O-methyl transferase by catechols and polyphenols. Biochem. Pharmacol. **22**, 247 (1973)
12. Baldessarini, R.J., Karobath, M.: Effects of L-dopa and L-3-O-methyl-dopa on uptake of [^{3}H]-L-methionine by synaptosomes. Neuropharmacology **11**, 715 (1972)
13. Baldessarini, R.J., Kopin, I.J.: Assay of tissue levels of S-adenosylmethionine. Analyt. Biochem. **6**, 289 (1963)
14. Baldessarini, R.J., Kopin, I.J.: S-adenosylmethionine in brain and other tissues. J. Neurochem. **13**, 769 (1966)
15. Ban, T.A.: On-going national collaborative studies in Canada: Niacin in the treatment of schizophrenics. Psychopharmacol. Bull. **5**, 5 (1969)
16. Banerjee, S.P., Snyder, S.H.: Methyltetrahydrofolic acid mediates N- and O-methylation of biogenic amines. Science **182**, 74 (1973)
17. Barchas, J.D., Elliott, G.R., DoAmaral, J., Erdelyi, E., O'Connor, S., Bowden, M., Brodie, H., Berger, P., Renson, J., Wyatt, R.J.: Tryptolines: Formation from tryptamines and 5-MTHF by human platelets. Arch. Gen. Psychiat. **31**, 862 (1974)
18. Bartholini, G., Pletscher, A.: Distribution and metabolism of L-3-O-methyl-Dopa in rats. Brit. J. Pharmacol. Chemother. **40**, 461 (1970)
19. Baudry, M., Chast, F., Schwartz, J.-C.: Studies on S-adenosyl-homocysteine inhibition of histamine transmethylation in brain. J. Neurochem. **20**, 13 (1973)
20. Berlet, H.H., Matsumoto, K., Pschedit, G.R., Spaide, J., Bull, C., Himwich, H.E.: Biochemical correlates of behavior in schizophrenic patients. Schizophrenic patients receiving tryptophan and methionine or methionine together with a monoamine oxidase inhibitor. Arch. gen. Psychiat. **13**, 521 (1965)
21. Brune, G.G., Himwich, H.E.: Effects of methionine-loading on the behavior of schizophrenic patients. J. nerv. ment. Dis. **134**, 447 (1962)
22. Brune, G.G., Himwich, H.E.: Biogenic amines and behavior in schizophrenic patients. In: Recent Advances in Biological Psychiatry. Wortis, J. (ed.) New York: Plenum 1963, Vol. V, p. 144
23. Buscaino, G.A., Spadetta, V., Carella, A.: Il test di metilazione nella schizofrenia, Acta neurol. (Napoli) **24**, 113 (1969)
24. Cantoni, G.L.: Methylation of nicotinamide with a soluble enzyme system from rat liver. J. biol. Chem. **189**, 203 (1951)
25. Carey, M.C., Donovan, D.E., Fitzgerald, O., McCauley, F.D.: Homocystinuria – I. A clinical and pathological study of nine subjects in six families. Amer. J. Med. **45**, 7 (1968)
26. Chalmers, J.P., Baldessarini, R.J., Wurtman, R.J.: Effects of L-dopa on norepinephrine metabolism in the brain. Proc. nat. Acad. Sci. (Wash.) **68**, 662 (1971)
27. Cohen, S.M., Nichols, A., Wyatt, R., Pollin, W.: The administration of methionine to chronic schizophrenic patients: a review of ten studies. Biol. Psychiat. **8**, 209 (1974)
28. Cohn, C.K., Dunner, D.L., Axelrod, J.: Reduced catechol-O-methyl transferase activity in red blood cells of women with primary affective disorders. Science **170**, 1323 (1970)
29. Coper, H., Deyhle, G., Fähndrich, C., Fähndrich, E., Rosenberg, L. Strauss, S., Blum, A., Dufour, H.: Excretion of vanillyl-mandelic acid, homovanillic acid, N-methyl-nicotinamide and N-methyl-2-pyridone-5-carboxamide in urine of voluntary test persons and psychiatric patients before and after administration of methionine. Pharmakopsychiat. Neuropsychopharmakol. **5**, 177 (1972)
30. Creveling, C.R., Daly, J.W.: Identification of 3,4-dimethoxy-phenethylamine from schizophrenic urine by mass spectrometry. Nature (Lond.) **216**, 190 (1967)
31. Deguchi, T., Barchas, J.: Inhibition of transmethylations of biogenic amines by S-adenosylhomocysteine. J. biol. Chem. **246**, 3175 (1971)
32. Domino, E.F., Krause, R.R., Bowers, J.: Various enzymes involved with putative neurotransmitters. Arch. gen. Psychiat. **29**, 195 (1973)

33. Donaldson, K.O., Keresztesy, J.C.: Further evidence on the nature of prefolic A. Biochem. biophys. Res. Commun. **5**, 289 (1961)
34. Dunner, D.L., Cohn, C.K., Weinshilboum, R.M., Wyatt, R.J.: The activity of dopamine-β-hydroxylase and methionine-activating enzyme in blood of schizophrenic patients. Biol. Psychiat. **6**, 215 (1973)
35. Fazio, C., Andreoli, V., Agnoli, A., Casacchia, M., Cerbo, R.: Effetti terapeutici della S-adenosil-metionina (SAMe) nelle sindromi depressive. Minerva med. (Roma) **64**, 1515 (1973)
36. Fischer, E., Spatz, H.: Studies on urinary elimination of bufotenine-like substances in schizophrenia. Biol. Psychiat. **2**, 235 (1970)
37. Friedhoff, A.J., Winkle, Van: Characteristics of an amine found in urine of schizophrenic patients. J. nerv. men. Dis. **135**, 550 (1962)
38. Gillin, J.C., Cannon, E., Magyar, R., Schwartz, M., Wyatt, R.J.: Failure of N, N-dimethyltryptamine to evoke tolerance in cats. Biol. Psychiat. **7**, 213 (1973)
39. Hall, P., Hartridge, G., Leeuwen, G.H., van: Effect of catechol-O-methyltransferase in schizophrenia. Arch. gen. Psychiat. **20**, 573 (1969)
40. Hartley, R., Padwick, D., Smith, J.A.: The inhibition of pineal hydroxyindole-O-methyltransferase by haloperidol and fluphenazine. J. Pharm. (Lond.) **24** (Suppl.), 100 (1972)
41. Haydu, G.G., Dhrymiotis, A., Korenyl, C., Goldschmidt, L.: Effects of methionine and hydroxychloroquine in schizophrenia. Amer. J. Psychiat. **122**, 560 (1965)
42. Heath, R.G., Nesselhof, W., Jr., Timmons, E.: D, L-methionine-d, l, -sulfoximine effects in schizophrenic patients. Arch. gen. Psychiat. **14**, 213 (1966)
43. Heslinga, F.J.M., Tilburg, W., van, Stam, F.C.: The identity of the so-called pink spot in schizophrenia and Parkinson's disease. Psychiat. Neurol. Neurochir. (Amst.) **73**, 157 (1970)
44. Hoffer, A., Osmond, H.: Treatment of schizophrenia with nicotinic acid: a ten-year follow-up. Acta psychiat. scand. **40**, 171 (1964)
45. Horwitt, M.K., Harvey, C.C., Rothwell, W.S., Cutter, J.L., Hoffron, D.: Tryptophanniacin relationship in man. J. Nutr. **60** (Suppl.), 1 (1956)
46. Hsu, L.L., Mandell, A.J.: Multiple N-methyltransferases for aromatic alkylamines in brain. Life Sci. **13**, 847 (1973)
47. Hsu, L.L., Mandell, A.J.: Enzymatic formation of tetrahydro-β-carboline from tryptamine and 5-methyltetrahydrofolic acid in rat brain fractions: regional and subcellular distribution. J. Neurochem. **24**, 631 (1975)
48. Israelstam, D.M., Sargent, T., Finley, N.N., Winchell, H.S., Fish, M.B., Motto, J., Pollycove, M., Johnson, A.: Abnormal methionine metabolism in schizophrenic and depressive states. J. psychiat. Res. **7**, 185 (1970)
49. Kakimoto, Y., Sano, I., Kanazawa, A., Tsujio, T., Kaneko, Z.: Metabolic effects of methionine in schizophrenic patients pretreated with a monoamine oxidase inhibitor. Nature (Lond.) **216**, 1110 (1967)
50. Kopin, I.J., Baldessarini, R.J.: Assay of tissue levels of S-adenosylmethionine. In: Methods in Enzymology, Vol. 17B: Metabolism of amino acids and amines. Tabor, H., Tabor, C.W. (eds.) New York: Academic Press, 1971, p. 397
51. Korevaar, W.C., Geyer, M.A., Knapp, S., Hsu, L.L., Mandell, A.J.: Regional distribution of 5-methyl-tetrahydrofolic acid in brain. Nature New Biol. **245**, 244 (1973)
52. Laduron, R.: N-methylation of dopamine to epinine in brain tissue using N-methyltetrahydrofolic acid as the methyl donor. Nature New Biol. **238**, 212 (1972)
53. Laduron, P.M., Gommersen, W.R., Leysen, J.E.: N-methylation of biogenic amines. Biochem. Pharmacol. **23**, 1599 (1974)
54. Lin, R.-L., Narasimhachari, N., Himwich, H.E.: Inhibition of indolethylamine-N-methyltransferase by S-adenosylhomocysteine. Biochem. biophys. Res. Commun. **54**, 751 (1973)
55. Lin, R.-L., Narasimhachari, N.: Evidence for the absence of amine-N-methylation and O-methylation in indolethylamines with methyltetrahydrofolic acid-dependent N-methyltransferase. Res. Commun. chem. Path. Pharmacol. **8**, 535 (1974)

56. Lipinski, J.F., Schaumburg, H.H., Baldessarini, R.J.: Regional distribution of histamine in human brain. Brain Res. **52**, 403 (1973)
57. Lipinski, J.F., Mandel, L.R., Ahn, H.S., Heuvel, W.J.A., van den, Walker, R.W.: Blood dimethyltryptamine concentrations in psychotic disorders. Biol. Psychiat. **9**, 89 (1974)
58. Lipton, M.A. (ed.): Megavitamin and orthomolecular therapy in psychiatry. American Psychiatric Association Task Force Report No. 7, Washington, D.C., 1973, pp. 54
59. Lombardini, J.B., Burch, M.K., Talalay, P.: An enzymatic derivative double-isotope assay for L-methionine. J. biol. Chem. **246**, 4465 (1971)
60. Mandel, L.R., Walker, R.W., Rosegay, A., Heuvel, W.J.A., van den, Rokach, J.: 5-methyltetrahydrofolic acid as a mediator in the formation of pyridoindoles. Science **186**, 741 (1974)
61. Mandell, A.J., Morgan, M.: Indole (ethyl) amine N-methyltransferase in human brain. Nature New Biol. **230**, 85 (1971)
62. Matthysse, S., Baldessarini, R.J.: S-adenosylmethionine and catechol-O-methyltransferase in schizophrenia. Amer. J. Psychiat. **128**, 1310 (1972)
63. Matthysse, S., Baldessarini, R.J., Vogt, M.: Methionine adenosyltransferase: a double-isotope derivative, enzymatic assay. Analyt. Biochem. **48**, 410 (1972)
64. Matthysse, S., Lipinski, J., Shih, V.: L-dopa and S-adenosylmethionine. Clin. chim. Acta **35**, 253 (1971)
65. McKensie, R.M., Gholson, R.K.: A simple assay for methionine adenosyltransferase using cation exchange paper and liquid scintillation spectrometry. Analyt. Biochem. **53**, 384 (1973)
66. Meller, E., Rosengarten, H., Friedhoff, A.J.: Conversion of $[C^{14}]$-S-adenosylmethionine to $[C^{14}]$-formaldehyde and condensation with indoleamines: a side-reaction in N-methyltransferase assay in blood. Life Sci. **14**, 2167 (1974)
67. Meller, E., Rosengarten, H., Friedhoff, A.J.: 5-methyltetrahydrofolic acid is not a methyl donor for biogenic amines: evidence for the enzymatic formation of formaldehyde, Science **187**, 171 (1975)
68. Murphy, D.L.: Mental effects of L-dopa. Ann. Rev. Med. **24**, 209 (1973)
69. Murphy, D.L., Wyatt, R.J.: Reduced monoamine oxidase activity in blood platelets from schizophrenic patients. Nature (Lond.) **238**, 225 (1972)
70. Narasimhachari, N., Himwich, H.E.: GC-MS identification of bufotenin in urine samples from patients with schizophrenia or infantile autism. Life Sci. II **12**, 475 (1973)
71. Narasimhachari, N., Himwich, H.E.: Gas chromatographic-mass spectrometric identification of N, N-dimethyltryptamine in urine samples from drug-free chronic schizophrenic patients and its quantitation by the technique of single (selective) ion monitoring. Biochem. biophys. Res. Commun. **55**, 1064 (1973)
72. Narasimhachari, N., Plaut, J., Himwich, H.E.: 3,4-Dimethoxyphenylethylamine, a normal or abnormal metabolite? J. psychiat. Res. **9**, 325 (1972)
73. Osmond, H., Smythies, J.: Schizophrenia: a new approach. J. ment. Sci. **98**, 309 (1952)
74. Park, L.C., Baldessarini, R.J., Kety, S.S.: Methionine effects in chronic schizophrenics. Arch. gen. Psychiat. **12**, 346 (1965)
75. Pollin, W., Cardon, P.V., Jr., Kety, S.S.: Effects of amino acid feedings in schizophrenic patients treated with iproniazid. Science **133**, 104 (1961)
76. Price, J.: Methylation in schizophrenics: a pharmacogenetic study. J. psychiat. Res. **9**, 345 (1972)
77. Rubin, R.A., Ordonez, L.A., Wurtman, R.J.: Physiological dependence of brain methionine and S-adenosylmethionine concentrations on serum amino acid pattern. J. Neurochem. **23**, 227 (1974)
78. Saavedra, J.M., Coyle, J.T., Axelrod, J.: The distribution and properties of the non-specific N-methyltransferase in brain. J. Neurochem. **20**, 743 (1973)

79. Salvatore, F., Utili, R., Zappia, V.: Quantitative analysis of S-adenosylmethionine and S-adenosylhomocysteine in animal tissues. Analyt. Biochem. **41**, 16 (1971)
80. Schwartz, M.A., Aikens, A.M., Wyatt, R.J.: Monoamine oxidase activity in brains from schizophrenic and mentally normal individuals. Psychopharmacologia (Berl.) **38**, 319 (1974)
81. Schweitzer, J.W., Friedhoff, A.J., Angrist, B.M., Gershon, S.: Excretion of *p*-methoxy-amphetamine administered to humans. Nature (Lond.) **229**, 133 (1971)
82. Sharma, S., Sinari, V.P.: Pink spot in the urine and C.S.F. of schizophrenics. Dis. nerv. Syst. **32**, 831 (1971)
83. Simpson, G.M., Varga, V.: An investigation of the clinical effect of GPA-1714, a catechol-O-methyltransferase inhibitor. J. clin. Pharmacol. **12**, 417 (1972)
84. Sloane, K.M., Bridges: The in vitro uptake of [^{35}S]-L-methionine by normal and leukaemic leucocytes. Acta haemat. (Basel) **40**, 18 (1968)
85. Snyder, S.H., Baldessarini, R.J., Axelrod, J.: A sensitive and specific enzymatic isotopic assay for tissue histamine. J. Pharmacol. exp. Ther. **153**, 544 (1966)
86. Spaide, J., Tanimukai, H., Bueno, J.R., Himwich, H.E.: Behavioral and biochemical alterations in schizophrenic patients. Arch. gen. Psychiat. **18**, 658 (1968)
87. Sprince, H.: An appraisal of methionine-tryptophan interrelationships in mental illness: methylation reactions involved. Biol. Psychiat. **2**, 109 (1970)
88. Sprince, H., Parker, C.M., Jameson, D., Elexander, F.: Urinary indoles in schizophrenic and psychoneurotic patients after administration of tranylcypromine (Parnate) and methionine or tryptophan. J. nerv. ment. Dis. **137**, 246 (1963)
89. Tanimukai, H., Ginther, R., Spaide, J., Bueno, J.R., Himwich, H.E.: Detection of psychotomimetic N, N-dimethylated indoleamines in the urine of four schizophrenic patients. Brit. J. Psychiat. **117**, 421 (1970)
90. Taylor, K.M., Snyder, S.H.: Isotopic microassay of histamine, histidine, histidine decarboxylase and histamine methyl transferase in brain tissue. J. Neurochem. **10**, 1343 (1972)
91. Tran-Manh, N., Laplante, M., Saint-Laurent, J., LeBel, E.: Abnormalities in $^{14}CO_2$ production from D.L-3,4-dihydroxyphenylalanine-[carboxyl-^{14}C] in schizophrenia and parkinsonism. Rev. canad. Biol. **31** (Suppl.), 255 (1972)
92. Wurtman, R.J., Rose, C.M., Matthysse, S., Stephenson, J., Baldessarini, R.J.: L-dihydroxyphenylalanine: effect on S-adenosylmethionine in brain. Science **169**, 395 (1970)
93. Wyatt, R.J., Mandel, L.R., Ahn, H.S., Walker, R.W., Heuvel, W.J.A., van den: Gas chromatographic-mass spectrometric isotope dilution determination of N, N-dimethyltryptamine concentrations in normals and psychiatric patients. Psychopharmacologia (Berl.) **31**, 265 (1973)
94. Wyatt, R.J., Murphy, D.L., Belmaker, R., Cohen, S., Donnelly, C.H., Pollin, W.: Reduced monoamine oxidase activity in platelets: a possible genetic marker to vulnerability to schizophrenia. Science **179**, 916 (1973)
95. Wyatt, R.J., Saavedra, J.M., Axelrod, J.: A dimethyltryptamine-forming enzyme in human blood. Amer. J. Psychiat. **130**, 754 (1973)
96. Wyatt, R.J., Erdelyi, E., Do Amaral, J.R., Elliott, G.R., Renson, J., Barchas, J.D.: Tetrahydro-β-carboline formation by human platelets and brain from methyltetrahydrofolic acid and tryptamine. Science. In press (1977)
97. Yaryura-Tobias, J.A., Merlis, S.: Levodopa and schizophrenia. J. Amer. med. Ass. **211**, 1857 (1970)
98. Zappia, V., Zydek-Cwick, C.R., Schlenck, F.: The specificity of S-adenosylmethionine derivatives in methyl transfer reactions. J. biol. Chem. **244**, 4499 (1969)

S-Adenosyl-L-Methionine (SAMe) and Biogenic Amine Methylation Processes

G. STRAMENTINOLI[1] and F. MAFFEI[2]

Among catecholamines, norepinephrine (NE) and dopamine (DA) serve as synaptic mediators in the central nervous system (CNS). The dopaminergic pathways constitute the ascending nigrostriatal system and connect the dorsal neurons of the interpeduncular nucleus with the accumbens nucleus and the olfactory tubercle (20). The noradrenergic pathways, all of them ascending, originate from neurons in the pons and medulla oblongata and spread partly to the hypothalamus, preoptic area, septum, amygdala, hippocampus, and cingulate gyrus, and partly to the neocortex.

Thus these catecholaminergic pathways are part of the limbic system, which controls affect and hence many aspects of behavior. That catecholamines function as mediators is confirmed by the observed fact that modifications of their turnover are associated with alterations of behavior (18).

In the biosynthesis of catecholamines (8, 9), DA is hydroxylated to NE, and this is then N-methylated to epinephrine.

The methyl group of epinephrine is supplied by S-adenosyl-L-methionine (SAMe) (14). This methylation reaction is catalyzed by phenylethanolamine-N-methyltransferase (PNMT), a soluble enzyme found in the adrenal glands and in some areas of the brain (hypothalamus, olfactory tubercle, and caudate nucleus) of several animal species (5, 17). Also in man, PNMT occurs at its highest concentrations in the adrenals and hypothalamus (15).

Quantitative and topographic assays of brain radioactivity by autoradiography in cats and rats treated with ^{14}C-methyl-labeled SAMe showed that tracer distribution favors the hippocampus, caudate nucleus, and hypothalamus: that is, the areas where PNMT activity and catecholamine concentrations are greatest. Also the naturally occurring SAMe in the CNS shows a characteristic topographic distribution with maximum concentrations in the basal nuclei (7).

SAMe as a methyl donor, however, is involved in the metabolism of catecholamines not only on the synthesis side but also on the catabolic side, where important transformations are governed mainly by the enzymes catechol-O-methyltransferase (COMT) and monoamine oxidase (MAO).

COMT enzymes, being responsible for the extraneuronal inactivation of catecholamines, require SAMe as a cofactor (5). In vivo, O-methylation occurs only in the meta position. In the rat, COMTs exhibit a characteristic distribution in the brain, with maxi-

---

[1] BioResearch Company, Department of Biochemistry, Liscate (Milan), Italy.

[2] Neuropsychiatric Hospitals, Verona (Italy).

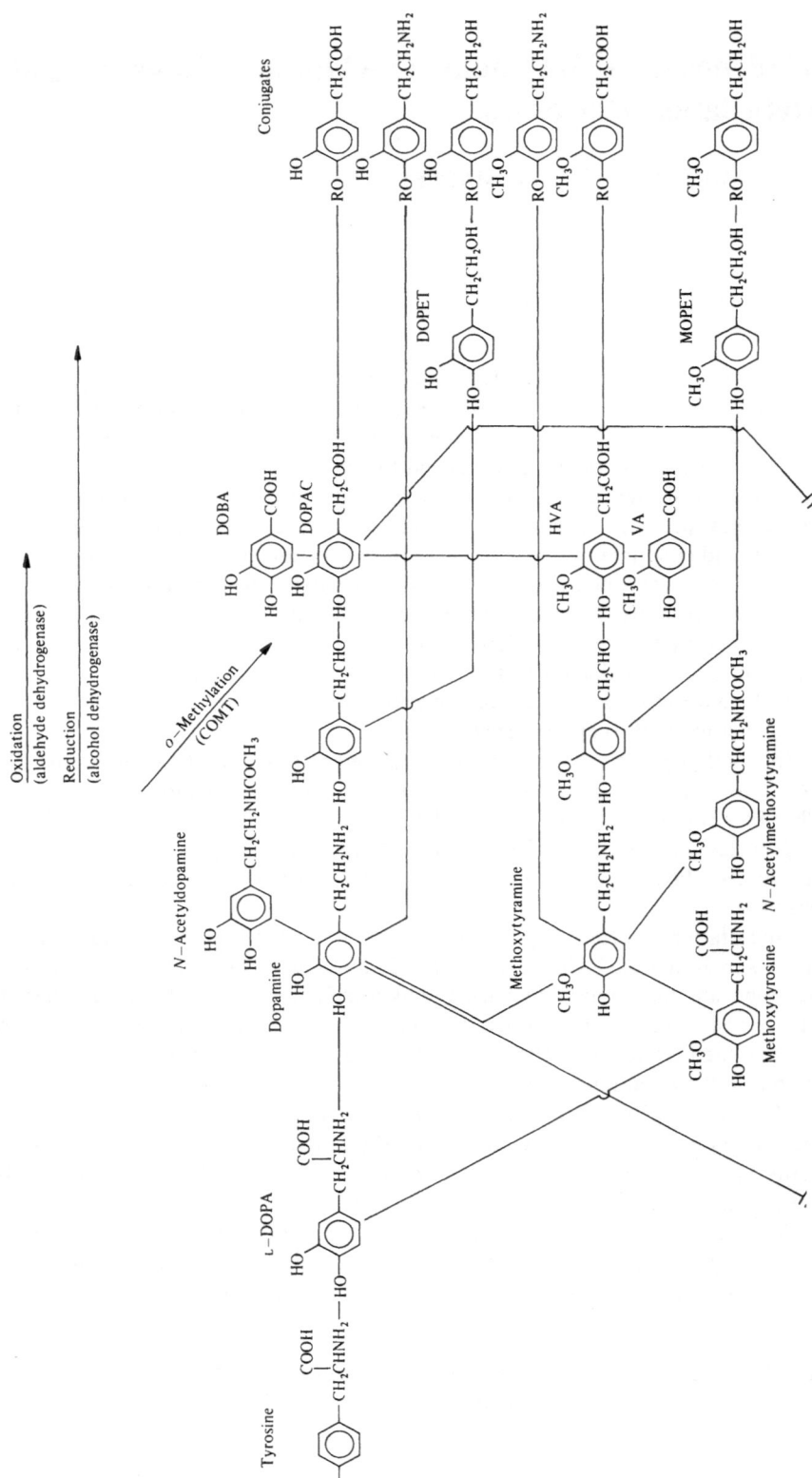

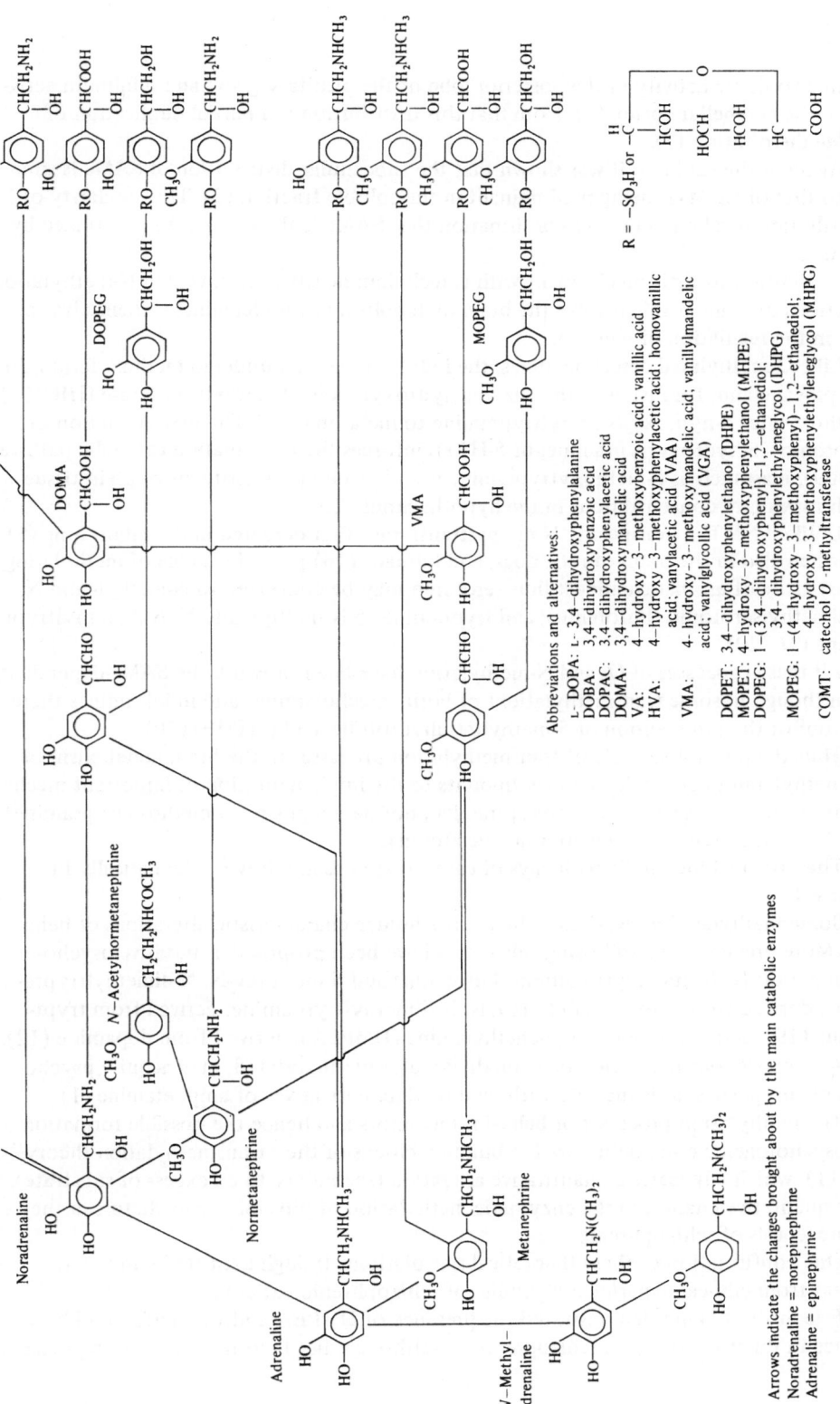

Abbreviations and alternatives:

L-DOPA: L- 3,4 –dihydroxyphenylalanine
DOBA: 3,4 –dihydroxybenzoic acid
DOPAC: 3,4 –dihydroxyphenylacetic acid
DOMA: 3,4 –dihydroxymandelic acid
VA: 4 –hydroxy –3 –methoxybenzoic acid; vanillic acid
HVA: 4 –hydroxy –3 –methoxyphenylacetic acid; homovanillic
         acid; vanilactic acid (VAA)
VMA: 4- hydroxy –3 –methoxymandelic acid; vanillylmandelic
         acid; vanylglycollic acid (VGA)

DOPET: 3,4 –dihydroxyphenylethanol (DHPE)
MOPET: 4-hydroxy –3 –methoxyphenylethanol (MHPE)
DOPEG: 1–(3,4 –dihydroxyphenyl)–1,2 –ethanediol;
         3,4- dihydroxyphenylethyleneglycol (DHPG)
MOPEG: 1–(4 –hydroxy –3 –methoxyphenyl)–1,2 –ethanediol;
         4 –hydroxy –3 –methoxyphenylethyleneglycol (MHPG)
COMT: catechol O -methyltransferase

Arrows indicate the changes brought about by the main catabolic enzymes
Noradrenaline = norepinephrine
Adrenaline = epinephrine

Fig. 1. The major changes in chemical structure that occur during the metabolism of the catecholamines

mum enzymatic activity in the posterior lobe of the pituitary gland and minimum activity in the cerebellar cortex (6). Note that this distribution is a partial duplication of SAMe distribution (7).

Again in the rat brain it was shown that the subcellular distribution of SAMe is similar to that of COMTs, being predominant in the soluble fraction (2). This similarity of distribution is taken as further confirmation that SAMe is the methyl donor utilized by COMTs.

In addition to their involvement with catecholamine substrates, O- and N-methylation reactions utilizing SAMe involve the brain metabolism of indoleamines — themselves active in the regulation of behavior.

Like O-methylated catecholamines, the indoleamines can undergo methoxylation. In the pineal gland, for instance, the enzyme hydroxyindole-O-methyltransferase (HIOMT) methoxylates 5-hydroxy-N-acetyltryptamine to melatonin (5). The methoxylation of serotonin (5-hydroxytryptamine, or 5-HT) reinforces the deamination catabolic pathway (MAO), inasmuch as 5-methoxytryptamine is a physiologic constituent of brain tissue with maximum concentrations in the hypothalamus (20).

Unlike PNMT, a nonspecific N-methyltransferase first detected in the rabbit lung (5), and then also in the brain of rats, dogs, and humans (16) proved capable of methylating a vast array of aromatic amines: thus serotonin may be converted to N-methyl- and N, N-dimethylserotonin (bufotenin); and tryptamine to N-methyl- and N, N-dimethyltryptamine (16, 20).

All these processes of O- and N-methylation have been shown to be SAMe-dependent, even though in some transmethylations of both catecholamines and indoleamines there is proof of the intervention of 5-methyltetrahydrofolic acid (MTHF) (20).

Thus the physiologic role of transmethylation processes in the brain metabolism of phenethylamines and indoleamines amounts to the implementation of important mechanisms of inactivation (e.g., methoxytyramine, normetanephrine, 5-methoxytryptamine) and biological synthesis (epinephrine, melatonin).

The principal metabolic pathways of catecholamines are shown schematically in Figure 1.

Some methylated derivatives of bioamines induce characteristic alterations of behavior. More precisely, the following substances have been proposed as putative psychotomimetics: N, N-dimethylserotonin (bufotenin) and 5-methoxy-N, N-dimethyltryptamine, derived from serotonin (11, 16); N, N-dimethyltryptamine, derived from tryptamine (19); and 3,4-dimethoxyphenethylamine (DMPEA), derived from dopamine (12).

Also exogenous molecules, once methylated or methoxylated, may acquire psychotogenic properties, as is the case with methoxylate derivatives of amphetamine (1).

The methylation processes of behavior mediators and hence the possible formation of psychotogenic compounds are the building blocks of the "transmethylation theory" (4, 11), which advocates a quantitative alteration (secondary to an excess of substrate), or a qualitative change, in the enzymatic methylation of physiologic mediators in the pathogenesis of schizophrenia.

One confirmation of these theoretical speculations, though admittedly indirect, comes from clinical-experimental studies of schizophrenic patients.

Friedhoff (11) has demonstrated the presence of DMPEA and/or a bufoteninlike substance in the urine of acute schizophrenics — although diet factors may be of importance in studies of this type and certainly limit their clinical and diagnostic value and even more so their pathogenetic significance.

Clinically, however, there is evidence of a close correlation between transmethylations and schizophrenic symptoms: an increase of methylation processes, brought about by the administration of methyl donors, indirectly with betaine and methionine (alone or associated with MAO inhibitors), or directly with SAMe itself (3, 10, 11, 13), results in the exacerbation of acute psychotic symptoms; and conversely a throttling of methylation processes, obtained by administering methyl acceptors such as nicotinamide (11), produces clinical improvement.

Contrariwise, in depressive syndromes, where again an alteration of catecholamine metabolism in the brain has been advocated (18), significant improvements have been reported with the administration of SAMe (10).

All these summarily presented facts lack, at the moment, a demonstrated concatenation; but they afford a glimpse of such ties based on the following points:

1. The role of SAMe in methylation processes is a physiologic event of brain biochemistry
2. Alteration of these processes, even on the same substrates, results in anomalous methylations
3. Methylated and methoxylated derivatives of biogenic amines have the ability to modify behavior
4. Such compounds, however, have not yet been conclusively demonstrated in spontaneous psychiatric pathology
5. The growing number of methylating enzymes detected in the CNS suggests the existence of metabolic passages or stages that in certain situations or in certain areas of the brain might give rise to methylated products with psychotogenic effects.

Thus we begin to arrange the tesserae of a mosaic which is still largely hypothetical but also coming into better and better focus in biochemical terms.

References

1. Andreoli, V.: Anfetamine, loro derivati e comportamento. In: Quaderni di Neuropsicofarmacologia. Andreoli, V., Del Mastro, S., (eds.), Milano: 1971
2. Andreoli, V., Maffei, F., Tonon, G.C.: The subcellular distribution of S-adenosyl-L-methionine (SAMe) in the rat brain. This volume, p. 178
3. Antun, F.T., Burnett, G.B., Cooper, A.J., Daly, R., Smythies, J.R., Zealley, A.K.: The effects of L-methionine (without MAOI) in schizophrenics. J. psychiat. Res. **8**, 63 (1971)
4. Antun, F., Eccleston, D., Smythies, J.R.: Transmethylation processes in schizophrenia. In: Brain Chemistry and Mental Disease. Ho, B.T., McIsaac, W.M. (eds.) New York: Plenum 1971, pp. 61-71
5. Axelrod, J.: Methyltransferase enzyme in the metabolism of physiologically active compounds and drugs. In: Handbook of Experimental Pharmacology. Eichler, O., Farah, H., Herken, H., Welch, A.D. (eds.) New York: 1971, Vol. XXVIII/2, p. 610
6. Axelrod, J., Tomchick, R.J.: Enzymatic O-methylation of epinephrine and other catechols. J. biol. Chem. **233**, 702 (1958)
7. Baldessarini, R.J., Kopin, I.J.: S-adenosylmethionine in brain and other tissues. J. Neurochem. **13**, 769 (1966)

8. Blaschko, H.: Formation of catechol amines in the animal body. Brit. med. Bull. **13**, 162 (1957)
9. Blaschko, H.: Catecholamine biosynthesis. Brit. med. Bull. **29**, 105 (1973)
10. Fazio, C., Andreoli, V., Agnoli, A., Casacchia, M., Cerbo, R., Pinzello, A.: Therapy of schizophrenia and depressive disorders with S-adenosyl-L-methionine. Intern. Res. Comm. System (IRCS), Clin. Pharmacol. Ther. **2**, 1015 (1974)
11. Friedhoff, A.J.: Biogenic amines and schizophrenia. In: Biological Psychiatry. Mendels, J. (ed.) New York: Wiley, 1973, pp. 113-129
12. Friedhoff, A.J., Schweitzer, J.W., Miller, J.: The enzymatic formation of 3,4-di-O-methylated dopamine metabolites by mammalian tissues. Res. Commun. Path. Pharmacol. **3**, 293 (1972)
13. Kakimoto Y., Sano, I., Kanazawa, A., Tsujio, T., Kaneko, Z.: Metabolic effects of methionine in schizophrenic patients pretreated with a monoamine oxidase inhibitor. Nature (Lond.) **216**, 1110 (1967)
14. Kirshner, N., McGoodall, C.: The formation of adrenaline from noradrenaline. Biochim. biophys. Acta **24**, 658 (1957)
15. Kitabchi, A.E., Williams, R.H.: Phenylethanolamine-N-methyltransferase in human adrenal gland. Biochim. Biophys. Acta **178**, 181 (1969)
16. Mandell, A.J., Segal, D.S.: The psychobiology of dopamine and the methylated indoleamines with particular reference to psychiatry. In: Biological Psychiatry. Mendels, J. (ed.) New York: Wiley, 1973, p. 89
17. McGeer, P.L., McGeer, E.G.: Formation of adrenaline by brain tissue. Biochem. biophys. Res. Commun. **17**, 502 (1964)
18. Mendels, J., Stinnett, J.L.: Biogenic amine metabolism. Depression and mania. In: Biological Psychiatry. Mendels, J. (ed.) New York: Wiley, 1973, pp. 35-64
19. Saavedra, J.M., Axelrod, J.: Psychotomimetic N-methylated tryptamine formation in brain in vivo and in vitro. Science **175**, 1365 (1972)
20. Snyder, S.H., Banerjee, S.P., Yamamura, H.I., Greenberg, D.: Drugs, neurotransmitters, and schizophrenia. Science **184**, 1243 (1974)

SAMe and Histamine

M. CASACCHIA[1], G. SQUITIERI[2], and A. AGNOLI[2]

The ralationships between SAMe and histamine are evident on at least one level: namely that SAMe serves as the methyl donor in the methyltransferase system that seems to represent the main catabolic pathway of endogenous brain histamine.

We must, however, admit from the start that our knowledge concerning brain histamine, its metabolism, and its role as a neuronal mediator, is only just beginning to shape up; and accordingly the meaning of relationships between SAMe-dependent transmethylations and histamine, and their potential pathophysiologic implications also remain largely obscure.

In this chapter we shall first of all describe, so far as present knowledge permits, the distribution and metabolism of brain histamine, with special emphasis on elements that suggest the role of the substance as a chemical mediator of brain functions.

Next, we will try to match this basic information with what we have learned in the vast field of transmethylations, and highlight the points where the match is more cogent.

We feel that the numerous implications emerging from this collation are bound to bring into better focus, biochemically as well as clinically, many ideas that are now mere hypotheses.

Brain Histamine Distribution

Histamine has been detected in the CNS of all animal species so far investigated, man being one. Its absolute mean concentration in the whole brain is definitely lower than that of other biogenic amines: in the rabbit it has been assessed at 53 ng/g (9).

Lipinski and his colleagues (27) made a map of histamine distribution in the human brain by a modification of Snyder's isotopic-enzymatic method. Their findings, obtained in autopsy specimens from 11 subjects dead from various causes (mean age 50.8 years), essentially confirm those of several earlier studies in animals.

The distribution of histamine in the brain is not homogeneous, a characteristic shared by the catecholamines, serotonin, and acetylcholine. Furthermore, the brain contains histamine in two distinct sites or compartments, namely the neurons and the mast cells.

Very high histamine concentrations occur in the hypophysis and pineal body — the only two structures in the brain that contain many mast cells. In the hypophysis there is

---

[1] First Department of Nervous and Mental Diseases, University of Rome (Prof. C. Fazio).
[2] Department of Nervous and Mental Diseases, University of L'Aquila (Prof. A. Agnoli).

a significant correlation between histamine concentration and mast cell count (1); but whereas the mast cell contingent almost certainly account for the bulk of histamine in the posterior pituitary and peduncle, this does not seem to be true in the anterior pituitary: compound 48/80, which liberates histamine contained in mast cells, reduces the histamine content of the posterior pituitary in the cat by about 60% and that of the peduncle by about 70%, but does not reduce the histamine content of the anterior pituitary (2). In effect, the histamine contained in the anterior pituitary is distributed differently from that of other brain structures, as is shown by the fact that the effects of reserpine, chlorpromazine, and iproniazid on anterior pituitary histamine differ from those of the same drugs anywhere else in the brain (2). As for the pineal body, there too we find a significant correlation between histamine concentration and mast cell count; but in a number of animal species those concentrations were not altered by the administration of octylamine, which is claimed to release histamine even from mast cells that resist the action of compound 48/80.

Among CNS structures that contain no mast cells in normal conditions, the hypothalamus shows the highest concentrations of histamine (1381.1 ng/g in the mamillary bodies of man); medium concentrations occur in the mesencephalon (substantia nigra, red nucleus, and colliculi) and in the olfactory bulb; definitely smaller amounts are found in the brain cortex and spinal cord; and the lowest concentrations of all occur in the cerebellar cortex (Table 1).

Histamine concentrations, however, vary also from one place to another within the hypothalamus. In the monkey, the highest concentrations occur in the mamillary bodies; slightly lower concentrations are detected in the ventromedial and supraoptic nuclei, and far smaller amounts are found in the infundibular regions (46).

Histamine Localization Within the Cell

The intracellular localization of histamine seems to reflect the typical distribution of mediators, at least a part of it being contained in the synaptosomes, i.e., in presynaptic terminals. Working with a centrifugation technique called "incomplete equilibrium sedimentation" on rat brain homogenates, Kunar, Snyder, et al. (23-25, 44) recently succeeded in separating different populations of synaptosomes, each containing a different neuronal mediator. In these experiments, histamine showed two peaks of sedimentation: one corresponding to the peaks of norepinephrine and GABA, and the other in a thinner region of the gradient. The former peak indicated a subcellular distribution similar to that of norepinephrine and GABA (note that the localization of histamine methyltransferase was also very nearly the same). As for the latter peak, this coincided exactly with the sedimentation profile of labeled exogenous histamine. This suggested that the second peak reflected the scattering of endogenous histamine during homogenization and its subsequent attachment to particulate material less dense than the majority of hypothalamic synaptosomes.

The only exception to this type in intracellular distribution is the predominantly intranuclear localization of histamine in the brain of neonate rats (30, 34). Between 5 and 10 days after birth, the rat shows a brain histamine content about five times

Table 1. Histamine distribution in the human brain: data from autopsy material of 11 subjects dead from various causes

Region	Histamine content (ng/g ± SEM)
Olfactory bulb	212.6 ± 20.6
Brain cortex:	
Upper frontal	118.9 ± 15.8
Precentral	109.7 ± 14.8
Postcentral	97.5 ± 12.1
Occipital	94.6 ± 13.3
Amygdala	94.8 ± 10.4
Hippocampus	108.7 ± 10.7
Basal nuclei:	
Globus pallidus	147.0 ± 11.0
Putamen	102.3 ± 14.0
Caudate nucleus	90.5 ± 12.3
Claustrum	148.1 ± 19.7
Internal capsule	163.1 ± 14.9
Cerebral peduncle	110.1 ± 13.5
Substantia nigra	214.4 ± 11.0
Red nucleus	225.4 ± 25.8
Thalamus:	
Anterior	138.8 ± 11.8
Dorsal	118.3 ± 7.2
Pulvinar	87.4 ± 8.9
Hypothalamus:	
Anterior	629.6 ± 79.5
Middle	1067.5 ± 164.1
Mamillary bodies	1381.1 ± 219.0
Upper colliculi	259.4 ± 16.7
Lower colliculi	320.2 ± 27.1
Pons:	
Dorsal	122.1 ± 22.9
Ventral	96.0 ± 5.5
Bulb:	
Dorsal	87.3 ± 10.2
Ventral	108.5 ± 16.8
Cerebellum:	
Cortex	39.9 ± 5.7
Dentate nucleus	121.5 ± 12.8

(From Lipinski et al., Ref. 27).

greater than that of the adult animal[3]; brain histamine assays begin to drop gradually at 17 days of age.

Between 3 and 14 days of neonatal life, 90 % of the total brain histamine is inside the nuclei; by age 21 days, this ratio is already down to 50 % (52)

These findings are viewed as an expression of active neuronal growth in the first days of life; at the same time, the data suggest that histamine might play a major role in the growth of the rat brain.

An apparent confirmation of this interpretation is the fact that these exceptionally high concentrations of histamine are found especially in telencephalic structures, which are precisely the regions that develop at the fastest rate in the first days of life — as opposed to diencephalic and rhombencephalic structures, which are philogenetically more ancient and so grow faster during fetal life (45).

Evaluation Methods

The first methods to be used in assessing brain histamine were biological assays based on the rapid contraction of the isolated guinea pig ileum in the presence of anticholinergic and antiserotonergic drugs (9). At that stage, one important source of error was the presence of interfering materials, especially slow-reacting substances, which at high concentrations produce rapid contractions of the isolated gut preparation hard to distinguish from that caused by histamine; another source of error was the presence of antagonistic substances.

This made it necessary to separate histamine from interfering substances, as can be done by ion-exchange chromatography (1). The acid extracts of brain tissue used in these methods contain also some methylhistamine; but the amounts are very small, and the activity of methylhistamine on the isolated ileum is only about 1/200 that of histamine.

A simpler and more specific approach to brain histamine assay is based on the condensation of histamine with o-phthaldialdehyde followed by fluorometric determination (42). Unfortunately, the butanol extraction previously used for removing interfering materials was good for practically all tissues except the brain (10) and blood (17). From the brain, currently used solvents extract not only histamine but also spermidine (22, 29), which is about 500 times as abundant as histamine and reacts with o-phthaldialdehyde to make a fluorophore with the same spectrum as histamine — which thus assays at 10 to 11 times its real value. Here, too, the error can be circumvented by using modified extraction and chromatography methods.

More recently, however, Snyder and his associates (43) developed an isotopic-enzymatic method based on the methylation of endogenous histamine by an exogenous histamine methyltransferase; the methyl donor is ^{14}C-labeled SAMe. With some further refinements (45), this method has achieved a sufficient degree of reliability.

---

[3] Also the SAMe content of the CNS varies with age, being greatest in the neonate animal and declining with growth (6).

Histamine Biogenesis

Histamine found in the living body is essentially of endogenous origin. Only negligible amounts are ingested with food and absorbed from the gut; and another tiny quota may be synthesized through decarboxylation of histidine by bacterial enzymes, and then absorbed into the blood.

In the CNS, histamine supplied via the blood stream is found only in a few places – mainly the anterior and posterior hypophysis, which are apparently unable to synthesize their own histamine (3, 49); conversely, the highest rate of histamine synthesis occurs in the hypothalamus (48).

Histamine is synthesized in the brain by the decarboxylation of histidine, which is an essential amino acid in the rat but can be synthesized from propionic acid in man (Fig. 1).

The decarboxylation of histidine involves two distinct enzymes, namely (a) aromatic L-amino acid decarboxylase, and (b) histidine decarboxylase. Both enzymes require pyridoxal phosphate as cofactor, but only (b) is specific for the synthesis of histamine, whereas (a) works equally well on other substrates, such as 5-hydroxytryptophan and dopa. The same enzyme, aromatic L-amino acid decarboxylase, is inhibited by a-methyldopa (28), which influences the synthesis of endogenous histamine only to a negligible extent. On the other hand a-hydrazino-histidine and NSD-1055, both specific inhibitors of histidine-decarboxylase in the hypothalamus (26, 32, 39), have no effects whatever on hypothalamic serotonin or dopamine concentrations.

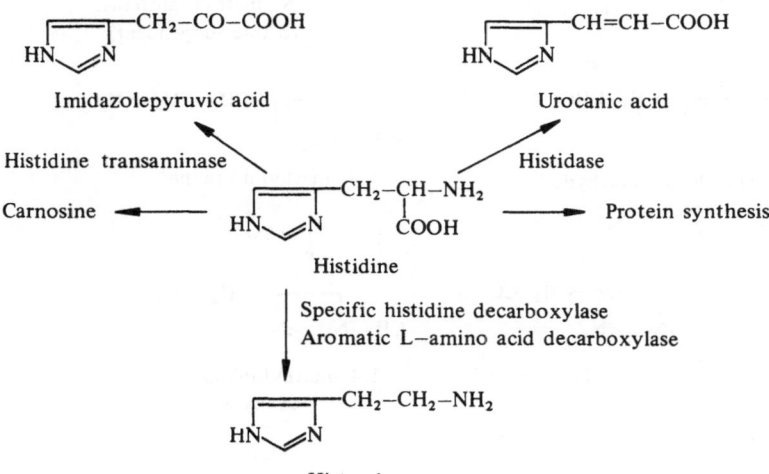

Fig. 1. Histamine biosynthesis

Histamine Catabolism

There are two main pathways of histamine degradation in brain tissue (Fig. 2):

(a) oxidative deamination
(b) N-methylation of the imidazole ring.

In the former instance, histamine produces imidazoleacetaldehyde by the action of a diamine oxidase which is inhibited by aminoguanidine (40), bulbocapnine (2), and imidazoleacetic acid (8). The resulting aldehyde is then very largely converted to imidazoleacetic acid by an aldehyde dehydrogenase.

In the latter instance, histamine is methylated by the agency of a histamine-N-methyltransferase to 1-methyl-4 (β-amino-ethylimidazole), better known as methylhistamine or 1,4-methylhistamine (as opposed to the 1,5-isomer). In this reaction, SAMe takes part as a methyl donor. Next, 1,4-methylhistamine undergoes oxidative deamination by an MAO which converts it to methylimidazoleacetaldehyde; and from there an aldehyde dehydrogenase produces methylimidazoleacetic acid, which is the principal urinary catabolite of histamine.

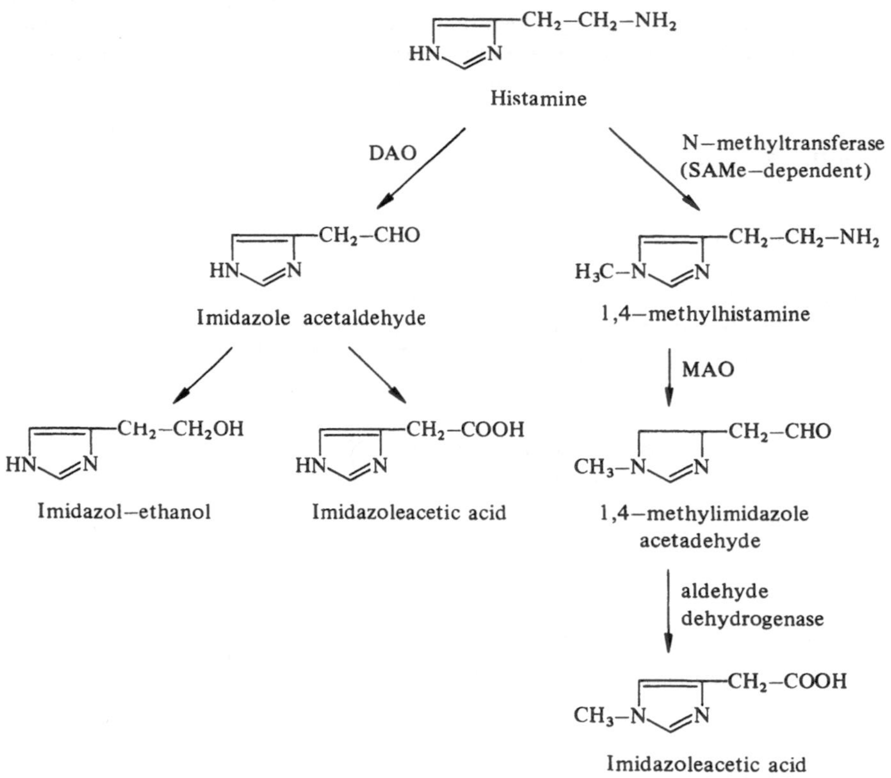

Fig. 2. Histamine catabolism. MAO = monoamine oxidase. DAO = diamine oxidase

The enzyme histamine methyltransferase is inhibited primarily by methylhistamine itself (33), and also by quinacrine, chlorpromazine, and $CuCl_2$ (21).

The function of the diamine oxidase system is almost certainly secondary and seems to be concerned mainly with the catabolism of exogenous histamine, albeit with the co-operation of the methyltransferase system. Endogenous histamine in the brain is catabolized exclusively, or nearly so, by transmethylation. In rats pretreated with aminoguanidine or methylhistamine by intracisternal injection, an intracerebral injection of ^{14}C-histamine results in a greater increase of brain radioactivity from labeled histamine than is the case with untreated control animals, the difference being more marked in rats pre-treated with methylhistamine. Conversely, in similarly pretreated rats, the concentration of ^{14}C-histamine synthesized from ^{14}C-L-histidine injected intracisternally is not modified by aminoguanidine and is significantly increased by methylhistamine (35-38).

Histamine methyltransferase is a highly specific enzyme with no activity on substrates other than histamine. Its distribution in the brain faithfully duplicates that of histamine, with peak concentrations in the hypothalamus, hypophysis, and pineal body, and lowest concentrations in the cerebral cortex and cerebellum (5) (Table 3). Recent studies (7, 31) indicate that with due allowance for the regional distribution of the two enzymes, histamine methyltransferase occurs in such brain cells as also contain COMTs — most probably glial cells.

Methylhistamine, in turn, is distributed in the brain essentially like histamine and histamine methyltransferase (51) (Table 4). In this respect, however, we must bear in mind that theoretically there is no need for a precise correlation between the brain distribution of methylhistamine and that of the transmethylating enzyme, since the actual concentrations of methylhistamine are determined also by the activity of other catabolic systems and of MAO enzymes.

As for the subcellular localization of methylhistamine, all we can say is that ultracentrifugation puts it in the crude mitochondrial fraction (13); currently available techniques do not afford greater resolution than that.

The distribution of methylhistamine in the brain has caused some authors (14, 16, 18, 50) to wonder if the substance was merely a catabolite of histamine or had a biological activity of its own.

Table 2. Inhibitors of the enzymatic metabolism of histamine

Histidine decarboxylase	NSD-1055
Aromatic L-amino acid decarboxylase	a-Hydrazino-histidine a-Methyldopa
Diamine oxidase	Aminoguanidine Bulbocapnine Imidazoleacetic acid
Histamine N-methyltransferase	Methylhistamine Chlorpromazine $CuCl_2$ Quinacrine
MAO	MAO inhibitors

Table 3. Distribution of histamine methyltransferase in the monkey brain (enzyme activity calculated as μmol of methylhistamine formed from histamine per g of tissue per h

Brain sites	Methylhistamine (μ mol/g/h)
Pons and bulb:	
Pons	0.41
Area postrema and dorsal bulb	054
Cerebellum (vermis)	0.36
Mesencephalon:	
Upper colliculi	0.49
Diencephalon:	
Dorsal thalamus	0.45
Hypothalamus	1.00
Habenula	0.61
Telencephalon:	
Cerebral cortex:	
Postcentral gyrus	0.57
Precentral gyrus	0.38
Caudate nucleus	0.62
Hippocampus	0.76
Olfactory bulb	0.78
Septum:	
Dorsal	0.97
Basal	0.66
Associated structures:	
Hypophysis:	
Anterior lobe	3.83
Posterior lobe	0.93
Pineal body	1.05

(Reduced, from Axelrod et al., Ref. 5).

Outside the CNS, mono- and dimethylated derivatives of histamine are known to have certain effects in the pancreas, digestive tract, lacrimal and salivary glands, and bronchial smooth musculature: more precisely, the actions is similar to that of histamine but less pronounced, with the exception of the gastric glands, where methylated histamine derivatives have been reported more active than histamine itself, at least in some animal species (8, 11). At the present state of our knowledge we cannot advocate an action of this kind for methylhistamine in the brain. What little evidence exists is conflicting. Hosli and his colleagues (20) tested the effects of histamine and its major catabolites on isolated neurons from the cat's brain stem, by a technique of micro-iontophoresis. Like histamine, methylhistamine was reported to have depressant effects on most of the neurons tested; this action, however, would be milder than that of GABA. We know from other quarters (16) that N, N-dimethylhistamine produces an awakening reaction in rabbits pretreated with pentobarbital. But perhaps these findings are only apparently at

Table 4. Methylhistamine distribution in the cat's brain

Brain sites	Methylhistamine (µg/g)
Cerebellum	0.1
Mesencephalon	0.27
Diencephalon:	
Thalamus	0.30
Hypothalamus	0.56
Telencephalon:	
Cerebral cortex	0.30
Caudate nucleus	0.40
Hypophysis	2.20

(From White, Ref. 51)

odds — maybe the correct interpretation is that the action of methylhistamine might be different or even opposite in different brain sites (presumably the brain stem and hypothalamus in the two examples just given[4].

Histamine Turnover

The turnover of histamine in the brain seems to be extremely rapid. Dismukes and Snyder (12) measured the half-life of histamine in the rat brain by two different methods:

(a) Intravenous injection of histidine decarboxylase inhibitors (bocresine and a-hydrazinohistidine). This caused a prompt fall (20-25%) of histamine assays in the hypothalamus and thalamomesencephalic region. Analysis of the depletion curve indicates a half-life of histamine of approximately 30.
(b) Analysis of the time pattern of appearance of ^{3}H-histamine in the brain after an intraventricular injection of ^{3}H-histidine. This gave a half-life value of about 1 min.

According to the authors just named, therefore, nearly all the histamine of the brain is contained in a rapid-turnover pool, very unlike the histamine contained in other body districts (the half-life of histamine in the gastic mucosa is about 2 h, and in mast-cell tissues it is up to 1 week); furthermore, the very short half-life of histamine in the brain also sets it apart from other biogenic amines: thus, for instance, brain dopamine, norepinephrine and serotonin have half-life values between 2 and 4 h — although dopamine and norepinephrine seem to occur also in small, rapid-turnover pools with half-life values between 5 and 10 min.

---

[4] Among other histamine catabolites, imidazoleacetic acid (and probably also methylimidazoleacetic acid) has a depressant action on brain neurons significantly greater than that of GABA itself. Green (18) has suggested that imidazoleacetic acid might have an important function in the CNS of lower animals; this agrees with the fact that diamine oxidase plays a leading role in phylologically inferior animal species.

The half-life of brain histamine, on the other hand, is very similar to that of acetylcholine.

This means that whereas the histamine concentrations in brain tissue are much lower than those of other neuronal mediators (with the exception of acetylcholine) in absolute values, from the dynamic point of view the number of histamine molecules synthesized per unit time is far greater than that of any other cerebral mediator (again, barring acetylcholine) (12).

Histamine as a Neuronal Mediator

As we have seen, therefore, there is enough evidence to place histamine among chemical mediators that are active in the CNS.

Its uneven distribution in the brain, with markedly selective peaks in certain areas (for instance the hypothalamic nuclei), repeats a characteristic of other biogenic amines and strongly suggests that histamine has some definite, specific functions that are mediated along equally definite histaminergic pathways.

Such pathways have not yet been identified with precision. Schwartz (41) reports that mechanical injuries to the diencephalon, involving the medial longitudinal bundle, produce a gradual decrement of histamine and L-histidine decarboxylase in the telencephalon of rats, the mode of this phenomenon suggesting anterograde degeneration of damaged histaminergic fibers. The same author advances the hypothesis of a specific histaminergic pathway emerging from the brain stem, crossing the lateral hypothalamus, and branching out to ipsilateral cortical areas. Unfortunately, this spatial arrangement agrees only partially with the specific distribution of histamine in the brain.

In subcellular terms, current fractionation techniques show that histamine is located preferentially in axonal endings; and it appears very probable that in such endings histamine concentrates within the presynaptic vesicles, as do the catecholamines.

Histamine may be liberated from nerve endings by depolarization of the brain tissue — again very strongly suggesting that the substance works as a neuronal mediator (45).

Also, wherever histamine is selectively localized in the brain, there we find also the specific metabolic systems that preside over its synthesis and degradation.

A further important fact is that the behavior of histamine in the brain is very different from that of the same substance in other body districts; in particular, the presence of a methylating system capable of inactivating the amine in a matter of seconds indicates that the function of histamine in the brain is different from that which occurs elsewhere in the body, where there is no such immediate interconnection between the anabolic and the catabolic sides of things (38).

Histamine and Psychotropic Drugs

Reserpine seems to have an action on histamine qualitatively similar to its action on the catecholamines and serotonin. Quantitatively, however, reserpine depletes histamine by liberating it from the presynaptic vesicles of axonal endings probably to a far lesser degree than it does the other biogenic amines. With reserpine dosages between 0.5 and 10 mg/kg, histamine assays in the hypothalamus showed a reduction of approximately 60% (2); in the mouse, maximal depletion in the whole brain was about 15% (4); and conversely in the rat no significant reduction was detected (19). However, even in such animal species as show the least apparent effects of reserpine in terms of histamine depletion (e.g., the rat), the metabolism of brain histamine does seem to be significantly affected (31). More precisely, the neosynthesis of histamine in the rat brain was reported increased by about 40% — an extent comparable to that of the increased synthesis of catecholamines and serotonin in the brain of rats treated with reserpine.

Also, whereas the rate of disappearance of radioactive histamine from the brain of rats treated with the labeled precursor and then with reserpine is significantly increased by the alkaloid, the concentration of methylhistamine shows only trifling changes certainly not commensurate with the assumed increase of histamine turnover. In other words, reserpine would impair the methylation process probably not by a mechanism of specific metabolic inhibition, but rather through a rerouting of histamine toward catabolic pathways other than methylation. This phenomenon has something in common with the behavior of catecholamines, which under the influence of reserpine are deaminated by intraneuronal MAO enzymes instead of making contact with extraneuronal COMTs.

This resemblance is the more striking if we recall that the enzyme histamine methyltransferase has the same apparent distribution as the COMTs' probably in glial cells. The two situations, on the other hand, differ in that histamine is not a substrate for MAO enzymes.

Chlorpromazine, administered three times daily to cats at a dosage of 50 mg/kg, increase histamine content in the hypothalamus by 50% (15); this action reflects the inhibition of histamine methyltransferase by chlorpromazine. Lower dosages of the drug produce much less increase of hypothalamic histamine, or none at all.

No comparable behavior is detected in the hypophysis, where histamine is trapped in the mast cells and apparently not available for methylation to any measurable extent.

Other MAO inhibitors such as pargyline (50 mg/kg), iproniazid (150 mg/kg), and tranylcypromine (25 mg/kg), administered acutely to animals, have no visible effects on histamine concentrations in the hypothalamus (45). Conversely, chronic administration of iproniazid (25 mg/kg for 5 days) seems to increase the histamine content of the hypothalamus and medial thalamus, by 46-74% and 28% respectively (18).

SAMe and Histamine

What we have said about relationships between SAMe and histamine, whether supported by experimental evidence or representing legitimate hypotheses, should by now appear at least to some extent evident. We must nevertheless bear in mind that the whole issue has begun to shape up only in the last few years — much too short a time for us to venture into final conclusions without fear of later contradiction. So at this point we shall only summarize the points that appear less assailable in the matter of relationships between SAMe and histamine in the CNS:

1. Transmethylation, in which SAMe is thought to play a crucial role as a donor of methyl groups, seems to represent the predominant pathway of catabolic degradation of endogenous brain histamine. On the other hand, what we have said about interactions between psychotropic drugs (e.g., reserpine and MAO inhibitors) and histamine suggests that transmethylation is certainly not the only a available pathway and perhaps not even the most important.

Another question that awaits settling is whether in the case of histamine transmethylation represents only a form of catabolic inactivation or whether it leads to the synthesis of biologically active substances whose action can be distinguished from that of histamine in terms either of quality or of magnitude.

2. With the exception of the cerebellum, there is at least a coarse correlation between the brain distribution of histamine and that of the methionine-activating enzyme (MAE) (47) (Table 5), which in turn reflects the distribution of SAMe.

Table 5. Brain distribution of methionine-activating enzyme (MAE)

Brain sites	Methionine-activating enzyme (units per g of tissue)
Hypothysis	142.0 ± 11.2
Hypothalamus	80.2 ± 3.2
Hippocampus	67.5 ± 3.0
Thalamus	59.8 ± 2.7
Parietal cortex	79.7 ± 2.6
Frontal cortex	72.6 ± 2.2
Parietal white matter	53.0 ± 2.8
Cerebellum	131.0 ± 4.8

(From J.J. Volpe et al., Ref. 47).

3. Because of its rapid turnover and high rate of synthesis, histamine is the greatest potential user SAMe among all cerebral amines. Consequently, the lack or deficiency of this methyl donor should result in immediate changes of histamine catabolism — namely an increased concentration of histamine in brain tissue, or perhaps the activation of alternative metabolic pathways; and in turn, pathologic situations reflecting a marked increase of endogenous histamine synthesis should interfere with the catabolism of other biogenic amines by subtracting SAMe from other enzyme systems such as catechol-O-methyltransferase (COMT), hydroxyindole-O-methyltransferase (HIOMT), and phenethanolamine-N-methyltransferase (PNMT).

References

1. Adam, H.M.: In: Regional Neurochemistry. Kety, S.S., Elkes, J., (eds.) London: Pergamon 1961, pp. 293-306
2. Adam, H.M., Hye, H.K.A.: Concentration of histamine in different parts of brain and hypophysis of cat and its modification by drugs. Brit. J. Pharmacol. **28**, 137 (1966)
3. Adam, H.M., Hye, H.K.A., Waton, N.G.: Studies on uptake and formation of histamine by hypophysis and hypothalamus in the cat. J. Physiol. (Lond.) **175**, 70 (1964)
4. Atack, C.: Reduction of histamine in mouse brain by N^1-(D, L-seryl)-N^2-(2, 3, 4-trihydroxybenzyl) hydrazine and reserpine. J. Pharmacol. **23**, 992 (1971)
5. Axelrod, J., MacLean, P., Albers Wayne, R., Weissbach, H.: In: Regional Neurochemistry. Kety, S.S., Elkes, J. (eds.) London: Pergamon 1961, pp. 307-311
6. Baldessarini, R.J., Kopin, I.J.: S-adenosylmethionine in brain and other tissues. J. Neurochem. **13**, 769 (1966)
7. Baudry, M., Chast, F., Schwartz, J.C.: Studies on S-adenosylhomocysteine inhibition of histamine transmethylation in brain. J. Neurochem. **20**, 13 (1973)
8. Bertaccini, G., Impiccitore, M., Mossini, F.: Action of some N-methyl derivatives of histamine on salivary and lacrimal secretion of the cat. Biochim. Pharmacol. **21**, 3076 (1972)
9. Carlini, E.A., Green, J.P.: The subcellular distribution of histamine, slow-reacting substance and 5-hydroxytryptamine in the brain of the rat. Brit. J. Pharmacol. **20**, 264 (1963)
10. Carlini, E.A., Green, J.P.: The measurement of histamine in brain and its distribution. Biochem. Pharmacol. **12**, 1448 (1963)
11. Code, C.F., Maslinski, S.M., Mossini, F., Navert, H.: Methyl histamines and gastric secretion. J. Physiol. (Lond.) **217**, 557 (1971)
12. Dismukes, K., Snyder, S.H.: Histamine turnover in rat brain. Brain Res. **78**, 467 (1974)
13. Fram, D.H., Green, J.P.: Methylhistamine in guinea-pig brain. J. Neurochem. **15**, 597 (1968)
14. Fram, D.H., Green, J.P.: Methylhistamine excretion during treatment with a monoamine oxidase inhibitor. Clin. Pharmacol. Ther. **9**, 355 (1968)
15. Furano, A.V., Green, J.P.: The uptake of biogenic amines by mast cells of the rat. J. Physiol. (Lond.) **170**, 263 (1964)
16. Goldstein, L., Pfeiffer, C.C., Munoz, C.: Quantitative EEG analysis of the stimulant properties of histamine and histamine derivatives. Fed. Proc. **22**, 424 (1963)
17. Graham, H., Scarpellini, J.A.D., Hubka, B.P., Lowry, O.H.: Measurement and normal range of free histamine in human blood plasma. Biochem. Pharmacol. **17**, 2271 (1968)
18. Green, J.P.: Histamine. In: Handbook of Neurochemistry Lajtha, A., (ed.) New York: Plenum, 1970, Vol. IV, pp. 222-250
19. Green, H., Frickson, R.W.: Effects of some drugs upon rat brain histamine content. Intern. J. Neuropharmacol. **3**, 315 (1964)
20. Hosli, L., Hass, A.L., Anderson, E.G.: The action of histamine and metabolites on single neurones of the mammalian central nervous system. J. de Pharmacologie, suppl. 1, Vol. 5, IX CINP Congress, Paris, 1974
21. Kahlson, G., Rosengren, E.: Biogenesis and physiology of histamine. Arnold, London: 1971
22. Kremzner, L.T., Pfeiffer, C.C.: Identification of substances interfering with fluorimetric determination of brain histamine. Biochem. Pharmacol. **14**, 1189 (1965)
23. Kunar, M.J., Green, A.I., Snyder, S.H., Gfeller, E.: Separation of synaptosomes storing catecholamines and GABA in rat corpus striatum. Brain Res. **21**, 405 (1970)

68

24. Kunar, M.J., Shaskan, E.G., Snyder, S.A.: The subcellular distribution of endogenous serotonin in brain tissue, comparison of synaptosomes storing serotonin, norepinephrine and GABA. J. Neurochem. **18**, 333 (1971)
25. Kunar, M.J., Taylor, K.M., Snyder, S.A.: The subcellular localization of histamine and histamine methyltransferase in rat brain. J. Neurochem. **18**, 1515 (1971).
26. Levine, R.J., Sato, T.L., Sjoerdsma, A.: Inhibition of histamine synthesis in the rat by α-hydrazino analog of histidine and 4-bromo-3-hydroxy benzyloxyamine. Biochem. Pharmacol. **14**, 139 (1965)
27. Lipinski, S.F., Schaumburg, H.H., Baldessarini, R.J.: Regional distribution of histamine in human brain. Brain Res. **52**, 403 (1973)
28. Lovenberg, W., Weissbach, H., Udenfriend, S.: Aromatic L-amino acid decarboxylase. J. biol. Chem. **237**, 89 (1962)
29. Michaelson, I.A., Coffman, P.Z.: An improved ion-exchange purification procedure for the fluorimetric assay of histamine. Analyt. Biochem. **27**, 257 (1969)
30. Pearce, L.A., Schanberg, S.M.: Histamine levels during brain development. Fed. Proc. **28**, 353 (1969)
31. Pollard, H., Bischoff, S., Schwartz, J.C.: Increased synthesis and release of H_3-histamine in rat brain by reserpine. Europ. J. Pharmacol. **24**, 399 (1973)
32. Reilly, M.A., Schayer, R.W.: Further studies on the histidine-histamine relationship in vivo: effects of endotoxin and of histidine decarboxylase inhibitors. Brit. J. Pharmacol. **34**, 551 (1968)
33. Reilly, M.A., Schayer, R.W.: In vivo studies on histamine catabolism and its inhibition. Brit. J. Pharmacol. **38**, 478 (1970)
34. Ronnberg, A.L., Schwartz, J.C.: Regional distribution of histamine in the brain of the rat. C.R. Acad. Sci. (Paris) **268**, 2376 (1969)
35. Schayer, R.W.: Determination of histidine decarboxylase activity. In: Analysis of Biogenic Amines and their Related Enzymes. Glick, D. (ed.) New York: Interscience, 1971, pp. 99-117
36. Schayer, K.W., Reilly, M.A.: In vivo formation and catabolism of (^{14}C)-histamine in mouse brain. J. Neurochem. **17**, 1649 (1970)
37. Schayer, R.W., Reilly, M.A.: Formation and fate of histamine in rat and mouse brain. J. Pharmacol. exp. Ther. **184**, 33 (1973)
38. Schayer, R.W., Reilly, M.A.: Metabolism of C_{14}-histamine in brain. J. Pharmacol. exp. Ther. **187**, 34 (1973)
39. Schayer, R.W., Yry, A.C.: Evidence that histamine is a gastric secretory hormone in the rat. Amer. J. Physiol. **189**, 369 (1957)
40. Schuler, W.: Zur Hemmung der Diaminooxydase (Histaminase). Experientia (Basel) **8**, 230 (1952)
41. Schwartz, J.C.: Histamine in an ascending pathway in brain: studies on localization, turnover and the effects of psychopharmacological agents. J. de Pharmacologie, suppl. 1, Vol. 5, IX CINP Congress, Paris, 1974
42. Shore, P.A., Burkhalter, A., Cohn, V.H.: A method for the fluorimetric assay of histamine in tissues. J. Pharmacol. exp. Ther. **127**, 182 (1959)
43. Snyder, S.H., Baldessarini, R.J., Axelrod, J.: A sensitive and specific enzymatic isotopic assay for tissue. J. Pharmacol. exp. Ther. **153**, 544 (1966)
44. Snyder, S.H., Glowinski, J., Axelrod, J.: The physiologic disposition of H_3-histamine in rat brain. J. Pharmacol. exp. Ther. **153**, 8 (1966)
45. Snyder, S.H., Taylor, K.M.: In: Perspectives in Neuropharmacology. Snyder, S.H. (ed.). London: Oxford University Press, 1972
46. Taylor, K.M., Geelier, E., Snyder, S.H.: Regional localization of histamine and histidine in the brain of the rhesus monkey. Brain Res. **41**, 171 (1972)
47. Volpe, J.J., Laster, L.: Trans-sulphuration in primate brain regional distribution of methionine activating enzyme in the brain of the Rhesus monkey at various stages of development. J. Neurochem. **17**, 413 (1970)

48. White, T.: Formation and catabolism of histamine in brain tissue in vitro. J. Physiol. (Lond.) **149**, 34 (1959)
49. White, T.: Formation and catabolism of histamine in cat brain in vivo. J. Physiol. (Lond.) **152**, 299 (1960)
50. White, T.: Some effects of histamine and two histamine metabolites on the cat's brain. J. Physiol. (Lond.) **159**, 198 (1961)
51. White, T.: Histamine and methylhistamine in cat brain and other tissues. Brit. J. Pharmacol. **26**, 494 (1966)
52. Young, A.B., Pert, C.D., Brown, D.G., Taylor, K.M., Snyder, S.H.: Nuclear localization of histamine in neonatal rat brain. Science **173**, 247 (1971)

S-Adenosyl-L-Methionine and GABA Metabolism in the Brain

A. CAROLEI[1], G. MECO[1], and A. AGNOLI[2]

Gamma-aminobutryric acid (GABA) (Fig. 1) and S-adenosyl-L-methionine (SAMe) have different and not completely understood roles within the framework of brain tissue metabolism.

$$H_2N-CH_2-CH_2-CH_2-COOH$$

Fig. 1. Gamma-aminobutyric acid (GABA)

S-adenosyl-L-methionine is involved in processes of transmethylation, whereas gamma-aminobutyric acid is regarded as an inhibitory mediator involved, among other things, in the regulation of intermediate brain metabolism (Fig. 2).

Now in the light of a careful review of GABA literature (36), we propose to explore all the possible functions of this substance and define possible correlations between such functions and transmethylation processes.

A good experimental model for the study of inhibitory neurotransmission has been the peripheral (neuromuscular) synapse of crustaceans. The work of Fatt and Katz (23) and of Dudel and Kuffler (20) has established two fundamental principles of inhibitory neurotransmission regulation, namely (1) that the inhibitory mediator may act by stabilizing the resting potential of the postsynaptic cell membrane by dint of producing selective changes of its ionic permeability; and (2), more directly, the mediator may act upon presynaptic (excitatory) nerve terminals by way of reducing the number of "quanta" of excitatory mediator that are liberated. Thus we have a postsynaptic action in the former case, and a presynaptic action in the latter.

Bazemore, Elliott, and Florey found that GABA extracted from the mammalian brain was most effective in inhibiting discharges into the stretch receptors of crustaceans (10); and Kuffler demonstrated its inhibitory effect on muscle, showing that GABA provided a close duplication of the action of the inhibitory neurotransmitter that occurs naturally in such organisms (44). Then Kravitz and his associates, working with an enzymatic fluorometric method, showed that GABA was more concentrated in inhibitory than in excitatory nerve fibers and co-workers (40, 42).

---

[1] First Department of Nervous and Mental Diseases, University of Rome (Prof. C. Fazio).
[2] Department of Nervous and Mental Diseases, University of L'Aquila (Prof. A. Agnoli).

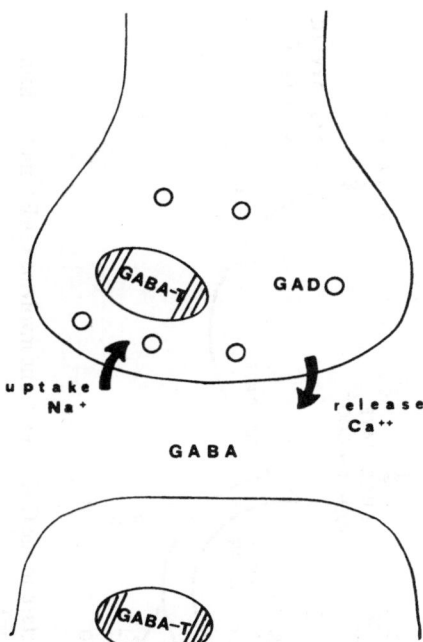

Fig. 2. Schematic representation of a GABA synapse. GABA-T = GABA-Glu transaminase. GAD = glutamic acid decarboxylase

   Again, Kravitz, Hall, and co-workers detected glutamic acid decarboxylase (GAD) only in inhibitory axons (35, 41).

   In mammalian organisms, Elliott and Jasper (1959) and Florey (1964) were the first to suggest that GABA and Factor I, an important inhibitory substance present in brain extracts, were one and the same thing (22, 25).

   Later on, Aprison, Davidoff, and Werman (1970) and Curtis (1969) demonstrated that at least in the spinal cord, glycine played a more important role than GABA (2, 14); but the role of the substance in the brain and cerebellum remained unassailable (43).

   More detailed studies of the distribution of GABA revealed that the substance occurred at high concentrations in the layer of Purkinje cells of the cerebellum and in the terminals of these neurons in the Deiters nucleus (45), as well as in the inhibitory interneurons of the hippocampus (27, 65).

   GABA is produced in nervous tissue by a decarboxylation reaction starting from L-glutamate (Fig. 3). Studies by various workers (38, 52, 60, 66), resulted in the isolation of the enzyme, L-glutamic acid decarboxylase (GAD I), catalyzing that reaction in the presence of pyridoxal phosphate. The enzyme was inhibited by chlorine ions (52, 66) and by pyridoxal antagonists, notably the "carbonyl trapping agents" (8).

   At this point a second enzyme (GAD II) was advocated, differing from GAD I by being insensitive to Cl$^-$ ions and by being stimulated rather than inhibited by the carbonyl trapping agents (32, 33). GAD II, however, does not give evidence of being implicated in the synthesis of GABA at inhibitory nerve end sites.

   GAD displays a conspicuous activity only in the axoplasm of inhibitory nerve terminals (35, 41), the synthesis being probably under the control of GABA (34, 68, 69) (Table 1).

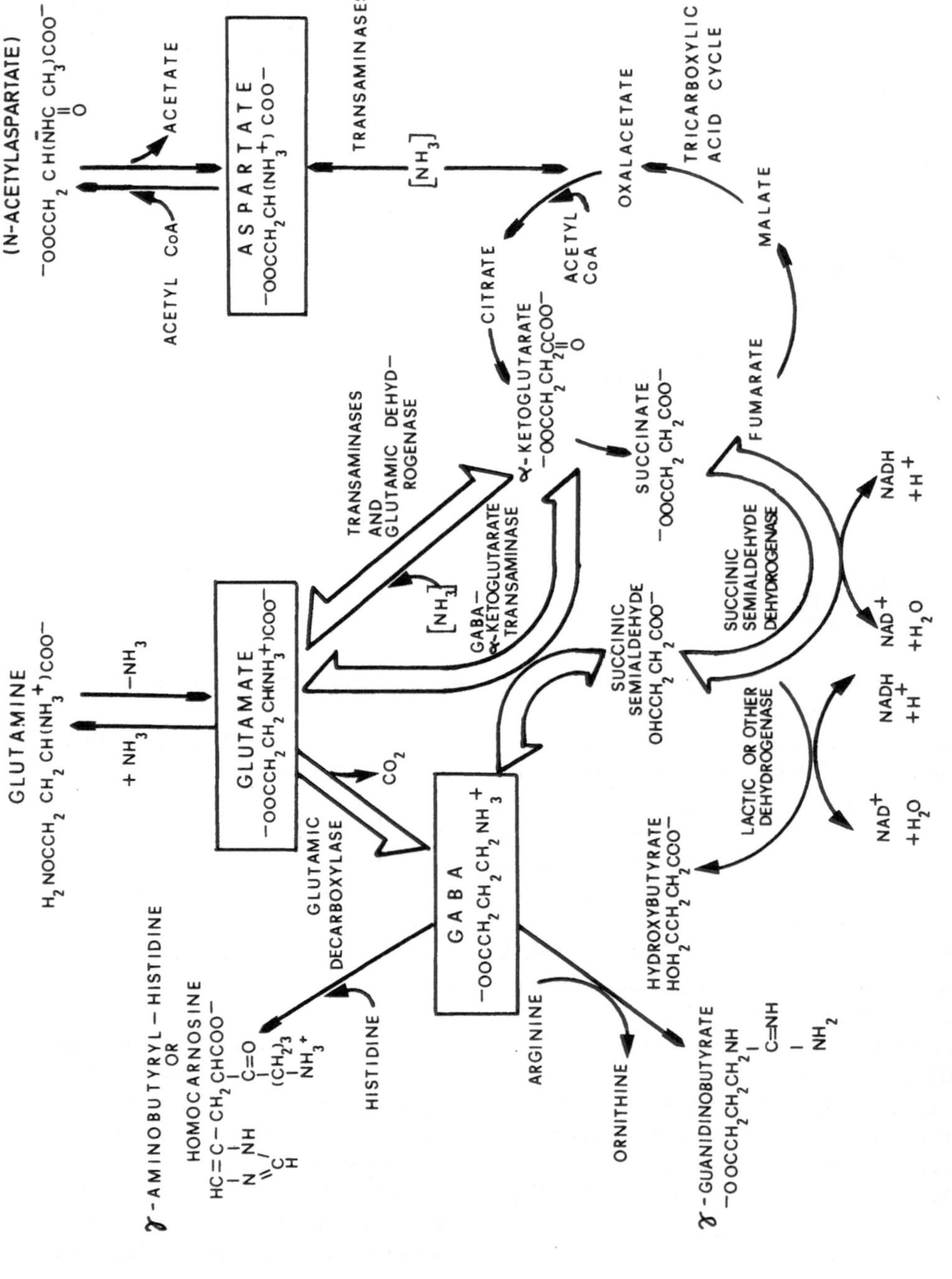

Fig. 3. The principal reactions of GABA, glutamate, and aspartate in the CNS. Reactions involving GABA metabolism are shown more prominently. (From Roberts, Ref. 59, redrawn)

Albers and Brady (1) detected significant regional differences of GAD activity in the Rhesus monkey brain, with high concentrations of the enzyme in the gray matter and low concentrations in the white matter; then Müller and Langeman confirmed similar differences in the human brain (53).

Table 1. GABA and related enzymes in the inhibitory (I) and excitatory (E) terminals of crustaceans (From Kravitz et al., Ref. 41; Hall et al., Ref. 35)

	I axons	E axons	I/E ratio
Axoplasm concentrations of GABA -M	0.105	0.0008	more than 100
Glutamate decarboxylase-p activity mol/cm/h	60	nil	more than 300
GABA-Glu transaminase-p activity mol/cm/h	32	23	1.5

GAD activity is especially strong in the Purkinje cells of the cerebellum (45); between 70 and 80% of that activity takes place in the synaptosomes (5, 26, 54, 62).

Both in crustaceans and in mammals the main pathway of GABA degradation involves its conversion to succinic semialdehyde. This reaction (incidentally, reversible) is catalyzed by pyridoxal-dependent GABA-T (GABA-Glu transaminase) (67, 72).

The next (and irreversible) passage to succinate is through an oxidation reaction that is catalyzed by SSAdh (succinic semialdehyde dehydrogenase) and NAD-dependent; this enzyme system is functionally synergistic with GABA-T (51, 64).

In the presence of NAD-dependent lactic acid dehydrogenase, the succinic semialdehyde may be converted to gamma-hydroxybutyric acid (24). This substance (and its immediate precursor, gamma-butyrolactone) has a marked CNS depressant action when administered to live animals; also, it increases striatal dopamine levels in the rat (47, 74).

Unlike GAD, GABA-T occurs also in other body districts, such as the liver and kidneys (13). Both GABA-T and SSAdh are found predominantly in the mitochondria, and their distribution in various brain locations is very much the same (35, 51). Only small amounts of both enzymes are found in synaptosome particle fractions, not only in GAB-aminergic neurons but also in those which do not utilize GABA as a neurotransmitter.

In the mammalian CNS, most of the endogenous GABA is accumulated in synaptic vesicles at nerve terminals, in places with a low affinity for the amino acid. Thus GABA tied up in the vesicles is at equilibrium with a plentiful fraction of free amino acids in the axoplasm (41).

GABA-inhibitory neurons are equipped with a specific uptake mechanism. The studies of Iversen and Neal (1968) and Iversen and Johnston (37) have demonstrated the high degree of specificity of the process.

At about the same time (1971), Bloom and Iversen were able to show that GABA uptake points were associated with inhibitory nerve endings in the mammalian CNS, by a method of autoradiography combined with the use of the electron microscope (12).

If you stimulate the inhibitory synaptic transmission you obtain an increased outflow of GABA (55). The amount of GABA liberated during nervous stimulation is in proportion to the number of stimuli, and requires the presence of calcium ions (55).

GAD activity is inhibited in vitro by hydrazide compounds that are convulsant in vivo; these seem to interfere with pyridoxal.

The hypothesis that convulsant hydrazides act by way of inhibiting GABA synthesis is corroborated by the finding of reduced GABA concentrations in the brain of animals treated with these compounds (6, 48).

Also, many carbonyl reagents such as hydroxylamine, cycloserine, and aminooxyacetic acid are potent inhibitors of GABA-T in vitro and in vivo (9, 17, 28, 73).

In evaluating these data we must not forget that GABA is the most potent inhibitory agonist of all, and that in crustaceans and mammals its effects are mediated by an increased permeability of postsynaptic membranes to chlorine ions.

The investigation of the main agonist and antagonist drugs with specific actions on GABA synapses was initiated by Edwards and Kuffler (21), Grundfest et al. (30), McGeer et al. (49), and Dudel (18, 19). The following are some highlights from such studies:

1. Bicuculline, an isoquinoline alkaloid, antagonizes GABA by structural competition; the effect is visible in the strychnine-insensitive neurons of the brain and cerebellar cortex and in the hippocampus (16, 56). Bicuculline also blocks the GAB-aminergic action of imidazoleacetic acid, being a product of histamine catabolism (31).
2. Picrotoxin blocks the pre- and postsynaptic actions of GABA in crustaceans (11, 25).
3. Semicarbazide depletes GABA in the CNS, the result being an increase of reflex responses (11).
4. Strychnine in high dosages antagonizes the actions of GABA on cortial neurons (39); and in low dosages it antagonizes synaptic inhibition by glycine (14). Strychnine, however, does not seem to depress the inhibitory effect of GABA on the central neurons of mammals (15).
5. Reserpine lowers GABA concentrations in the brain (7).
6. MAO inhibitors reduce the brain GABA content, probably by inhibiting the enzyme GABA-T (58).
7. Aminooxyacetic acid increases the GABA content by inhibiting the enzyme GABA-T (50).
8. Diphenylhydantoin, phenobarbitone and sodium dipropylacetate (all of them anticonvulsant drugs) increase the central levels of GABA (29, 61).

On the whole, all the compounds that interfere with the in vivo catabolism of GABA produce marked increases of the stationary GABA pool in the mammalian CNS. It is not entirely clear, however, whether these changes take place only at the sites of neuronal accumulation or whether they also modify the amounts of GABA that are liberated at inhibitory synapse terminals. Many of the componds just listed, incidentally, do not affect GABA metabolism specifically but have several other actions on CNS metabolism.

Now, knowledge so far accumulated in these matters does not indicate definite relationships between the biochemistry of SAMe and that of GABA. Seiler et al. (63) believe that putrescine (diaminobutane), a product of intestinal fermentation found also in the brain, can be deaminated by oxidation to gamma-aminobutyric aldehyde, which in turn could be oxidized to GABA. If confirmed, this finding would acquire considerable significance, suggesting that some of the brain GABA might originate from polyamines, without the intervention of glutamic acid (57).

The release of the aminopropyl group (71), effected by decarboxylated adenosine-methionine, seems to take place within the nuclei of nerve cells.

Recent work by Yessaian and Arakelian (75) shows that intraperitoneal injection of GABA lowers the concentration of norpinephrine in the brain of rats.

GABA appears to act by way of liberating norepinephrine from nerve terminals. The liberated norepinephrine is then converted to normetanephrine by extracellular COMTs and SAMe (3,4). Thus, indirectly, GABA seems to have the ability to lower SAMe concentrations in the brain.

All this evidence, however interesting, is of course in need of further and more detailed verification.

References

1. Albers, R.W., Brady, R.O.: The distribution of glutamic decarboxylase in the nervous system of the Rhesus monkey. J. biol. Chem. **234**, 926 (1959)
2. Aprison, M.H., Davidoff, R.A., Werman, R.: In: Handbook of Neurochemistry. Lajthe, A. (ed.) New York: Plenum, 1970, Vol. III, p. 381
3. Axelrod, J.: O-methylation of epinephrine and other catechols in vitro and in vivo. Science **126**, 400 (1957)
4. Axelrod, J., Senoh, S., Witkop, B.: O-methylation of catechol amines in vivo. J. biol. Chem. **233**, 697 (1958)
5. Balazs, R., Dahl, D., Harwood, J.R.: Subcellular distribution of enzymes of glutamine metabolism in rat brain. J. Neurochem. **13**, 897 (1966)
6. Balzer, H., Holtz, P., Palm, D.: Untersuchungen über die biochemischen Grundlagen der konvulsiven Wirkung von Hydraziden. Naunyn-Schmiedeberg's Arch. exp. Pharmakol. **239**, 520 (1960)
7. Balzer, H., Holtz, P., Palm, D.: Reserpin and Gamma-Aminobuttersäuregehalt des Gehirns. Experientia (Basel) **17**, 38 (1961)
8. Baxter, C.F.: In: Handbook of Neurochemistry. Lajthe, A. (ed.) New York: Plenum, 1970, Vol. III, p. 289
9. Baxter, C.F., Roberts, E.: Elevation of gamma-aminobutyric acid in rat brain with hydroxylamine. Proc. Soc. exp. Biol. (N. 4) **101**, 811 (1959)
10. Bazemore, A., Elliot, K.A.C., Florey, E.: Isolation of Factor I. J. Neurochem. **1**, 334 (1957)
11. Brill, J.A., Anderson, E.G.: The influence of semicarbazideinduced depletion of gamma-aminobutyric acid on presynaptic inhibition. Brain Res. **43**, 161 (1972)
12. Bloom, F.E., Iversen, L.L.: Localizing ^{3}H-GABA in nerve terminals of rat cerebral cortex by electron microscopic autoradiography. Nature (Lond.) **229**, 628 (1971)
13. Cacioppo, F., Pandolfo, L., Di Chiara, C.: Transaminazione acido gamma-aminobutirrico − acido alfa-chetoglutarico in alcuni tessuti di ratto. Boll. Soc. ital. Sper. **36**, 465 (1959)
14. Curtis, D.R.: The pharmacology of spinal postsynaptic inhibition. Prog. Brain Res. **31**, 171 (1969)
15. Curtis, D.R., Duggan, A.W., Felix, D., Johnston, G.A.R.: GABA, bicuculline and centra inhibition. Nature (Lond.) **226**, 1222 (1970)
16. Curtis, D.R., Felix, D., McLennan, H.: GABA and hippocampal inhibition. Brit. J. Pharmacol. **40**, 881 (1970)
17. Dann, O.T., Carter, C.E.: Cycloserine inhibition of gamma-aminobutyric-alpha-ketoglutaric transaminase. Biochem. Pharmacol. **13**, 677 (1964)
18. Dudel, J.: Presynaptic and postsynaptic effects of inhibitory drugs on the crayfish neuromuscular junction. Pflugers Archiv ges. Physiol. **104** (1965)
19. Dudel, J.: The action of inhibitory drugs on nerve terminals in cryfish muscle. Pflügers Archiv ges. Physiol. **284**, 81 (1965)
20. Dudel, J., Kuffler, S.W.: The quantal nature of transmission and spontaneous miniature potentials at the crayfish neuromuscular junction. J. Physiol. (Lond.) **155**, 534 (1961)
21. Edwards, C., Kuffler, S.W.: The blocking effect of gamma-aminobutyric acid (GABA) and the action of related compounds on single nerve cells. J. Neurochem. **4**, 19 (1959)

22. Elliot, K.A.C., Jasper, H.H.: Gamma-aminobutyric acid. Physiol. Rev. **39**, 383 (1959)
23. Fatt, P., Katz, B.: The effect of inhibitory nerve impulses on a crustacean muscle fiber. J. Physiol. (Lond.) **121**, 374 (1953)
24. Fishbein, W.N., Bessman, S.P.: Gamma-hydroxybutyrate in mammalian brain. Reversible oxidation by lactic dehydrogenase. J. biol. Chem **239**, 357 (1964)
25. Florey, E.: Further evidence for the transmitter-function of Factor I. Naturwissenschaften **44**, 424 (1957)
26. Fonnum, F.: The distribution of glutamate decarboxylase and aspartate transaminase in subcellular fractions of rat and guinea pig brain. Biochem. J. **106**, 401 (1968)
27. Fonnum, J., Storm-Mathisen, J.: GABA synthesis in rat hippocampus correlated to the distribution of inhibitory neurons. Acta physiol. scand. **76**, 35 A (1969)
28. Gelder, N.M., van: The effect of aminooxyacetic acid on the metabolism of gamma-aminobutyric acid in brain. Biochem. Pharmacol. **15**, 533 (1966)
29. Godin, Y., Heiner, L., Mark, J., Mandel, P.: Effects of di-n-propylacetate, an anticonvulsive compound, on GABA metabolism. J. Neurochem. **16**, 869 (1969)
30. Grundfest, J., Reube, J.P., Rickles, W.H.: The electrophysiology and pharmacology of lobster neuromuscular synapses. J. gen. Physiol. **42**, 1301 (1959)
31. Haas, H.L., Anderson, E.G., Hosli, L.: Histamine and metabolites: their effects and interactions with convulsants on brain system neurones. Brain Res. **51**, 269 (1973)
32. Haber, B., Kuriyama, K., Roberts, E.: An anion stimulated L-glutamic decarboxylase in nonneural tissues. Occurrence and subcellular localisation in mouse kidney and developing chick embryo brain. Biochem. Pharmacol. **19**, 1119 (1970)
33. Haber, B., Kuriyama, K., Roberts, E.: L-Glutamic acid decarboxylase: a new type in glial cells and human brain gliomas. Science **168**, 598 (1970)
34. Haber, B., Sze, P.Y., Kuriyama, K., Roberts, E.: GABA as a repressor of L-glutamic acid decarboxylase (GAD) in developing chick embryo optic lobes. Brain Res. **18**, 545 (1970)
35. Hall, Z., Bownds, M.D., Kravitz, E.A.: The metabolism of gamma-aminobutyric acid in the lobster nervous system. J. Cell Biol. **46**, 290 (1970)
36. Iversen, L.L.: The uptake, storage, release and metabolism of GABA in inhibitory nerves. In: Perspectives in Neuropharmacology. Snyder, S.H. (ed.) New York: Oxford University Press, 1972
37. Iversen, L.L., Johnston, G.A.R.: GABA uptake in rat central nervous system: comparison of uptake in slices and homogenates and the effects of some inhibitors. J. Neurochem. **18**, 1939 (1971)
38. Jenny, E., Solberg, R.: Biochemische Eigenschaften partiell gereinigter Glutaminsäure-decarboxylase aus Kalbshirnrinde. Helv. physiol. pharmacol. Acta **26**, 270 (1968)
39. Johnson, E.S., Roberts, M.H.T., Straughamn, D.W.: Amino-acid induced depression of cortical neurones. Brit. J. Pharmacol. **38**, 659 (1970)
40. Kravitz, E.A., Kuffler, S.W., Potter, D.D., Gelder, N.M., van: Gamma-aminobutyric acid and other blocking compounds in crustacea. II) Peripheral nervous system. J. Neurophysiol. **26**, 729 (1963)
41. Kravitz, E.A., Molinoff, P.B., Hall, Z.W.: A comparison of the enzymes and substrates of gamma-aminobutyric acid metabolism in lobster excitatory and inhibitory axons. Proc. nat. Acad. Sci. (Wash.) **54**, 778 (1965)
42. Kravitz, E.A., Potter, D.D.: A further study of the distribution of gamma-aminobutyric acid between excitatory and inhibitory axons of the lobster. J. Neurochem. **12**, 323 (1965)
43. Krnjevic, K.: Glutamate and gamma-aminobutyric acid in brain. Nature (Lond.) **228**, 119 (1970)
44. Kuffler, S.W.: Excitation and inhibition in single nerve cells. Harvey Lect., 1958/1959, 176, 1960
45. Kuriyama, K., Haber, B., Sisken, B., Roberts, E.: The gamma-aminobutyric acid system in rabbit cerebellum. Proc. nat. Acad. Sci. (Wash.) **55**, 846 (1966)

77

46. Kuriyama, K., Roberts, E., Rubinstein, M.K.: Elevation of gamma-aminobutyric acid in brain with amino-oxyacetic acid and susceptibility to convulsive seizures in mice: a quantitative re-evaluation. Biochem. Pharmacol. **15**, 221 (1966)
47. Laborit, H.: Sodium 4-hydroxybutyrate. Intern. J. Neuropharmacol. **3**, 433 (1964)
48. Maynert, E.W., Kaji, H.K.: On the relationship of brain gamma-aminobutyric acid to convulsions. J. Pharmacol. exp. Ther. **137**, 114 (1962)
49. McGeer, E.G., McGeer, P.L., McLennan, H.: The inhibitory action of 3-hydroxytyramine, gamma-aminobutyric acid (GABA) and some other compounds towards the crayfish stretch receptor neuron. J. Neurochem. **8**, 36 (1961)
50. Meldrum, B.S., Balzano, E., Gadea, M., Naquet, R.: Photic and drug-induced epilepsy in the baboon (*Papio-papio*): the effects of isoniazid, thiosemicarbazide, pyridoxine and amino-oxyacetic acid. Electroenceph. clin. Neurophysiol. **29**, 333 (1970)
51. Miller, A.L., Pitts, F.N.: Brain succinate semialdehyde dehydrogenase. III) Activities in twenty-four regions of human brain. J. Neurochem. **14**, 579 (1967)
52. Molinoff, P.B., Kravitz, E.A.: The metabolism of gamma-aminobutyric acid (GABA) in the lobster nervous system. Glutamic decarboxylase. J. Neurochem. **15**, 391, (1968)
53. Müller, P.B., Langemann, H.: Distribution of glutamic acid decarboxylase activity in human brain. J. Neurochem. **9**, 399 (1962)
54. Neal, M.J., Iversen, L.L.: Subcellular distribution of endogenous and ^{3}H gamma-aminobutyric acid in rat cerebral cortex. J. Neurochem. **16**, 1245 (1969)
55. Otsuka, M., Iversen, L.L., Hall, Z.W., Kravitz, E.A.: Release of gamma-aminobutyric acid from inhibitory nerves of lobster. Proc. nat. Acad. Sci. (Wash.) **56**, 1110 (1966)
56. Peck, E.J., Scaeffer, J.M., Clark, J.M.: Gamma-aminobutyric acid, bicuculline and postsynaptic binding sites. Biochem. biophys. Res. Commun. **52**, 394 (1973)
57. Pegg, A.E.: Biosynthesis of putrescine and polyamines in mammalian tissues. Ann. N.Y. Acad. Sci. **171**, 977 (1970)
58. Popov, N., Pohle, W., Matthies, H.: Einfluß von Phenelzin und Aminooxygessigsäure auf die gamma-Aminobuttersäure, Glutaminsäure-Decarboxylase und gamma-Aminobuttersäure-alpha-ketoglutarsäure-transaminase in verschiedenen Regionen des Rattenhirns. Acta. biol. med. germ. **20**, 509 (1968)
59. Roberts, E.: Gamma-aminobutyric acid and nervous system function. A perspective. Biochem. Pharmacol. **23**, 2637 (1974)
60. Roberts, E., Kuriyama, K.: Biochemical-physiological correlations in studies of the gamma-aminobutyric acid system. Brain Res. **8**, 1 (1968)
61. Saad, S., El Masry, A.M., Scott, P.M.: Influence of certain anticonvulsants on the concentration of gamma-aminobutyric acid in the cerebral hemispheres of mice. Europ. J. Pharmacol. **17**, 386 (1972)
62. Salganicoff, L., De Robertis, E.: Subcellular distribution of the enzyme of the glutamic acid, glutamine and gamma-aminobutyric acid cycles in rat brain. J. Neurochem. **12**, 287 (1965)
63. Seiler, N., Al-Therib, M.J., Kataoka, K.: Formation of GABA from putrescine in the brain of fish (*Salmo iridens gibb.*). J. Neurochem. **20**, 699 (1973)
64. Sheridan, J.J., Sims, K.L., Pitts, F.N.: Brain gamma-aminobutyrate-alpha-oxoglutarate transaminase. II) Activities in twenty-four regions of human brain. J. Neurochem. **14**, 571 (1967)
65. Storm-Mathisen, J., Fonnum, F.: Neurotransmitter synthesis in excitatory and inhibitory synapses of rat hippocampus. In: Second International Meeting of the International Society for Neurochemistry. Paoletti, R., Fumagalli, R., Galli, C. (eds.) Milano: Tamburini 1969, p. 382
66. Susz, J.P., Haber, B., Roberts, E.: Purification and some properties of mouse brain L-glutamic decarboxylase. Biochemistry **5**, 2870 (1966)
67. Sytinsky, I.A., Vasilijev, V.Y.: Some catalytic properties of purified gamma-aminobutyrate-alpha-oxoglutarate transaminase from the rat brain. Enzymologia **39**, 1 (1970)

78

68. Sze, P.Y.: Possible repression of L-glutamic acid decarboxylase by gamma-amino-butyric acid in developing mouse brain. Brain Res. **19**, 322 (1970)
69. Sze, P.Y., Lovell, R.A.: A reexamination of the effect of thiosemicarbazide on brain GABA and glutamic decarboxylase in vivo. Life Sci. **9**, 889 (1970)
70. Sze, P.Y., Lovell, R.A.: Reduction of level of L-glutamic acid decarboxylase by gamma-aminobutyric acid in mouse brain. J. Neurochem. **17**, 1657 (1970)
71. Tabor, H., Tabor, C.W.: Spermidine, spermine and related amines. Pharmacol. Rev. **16**, 245 (1964)
72. Waksman, A., Roberts, E.: Purification and some properties of mouse brain gamma-aminobutyric-alpha-ketoglutaric acid transaminase. Biochemistry **4**, 2132 (1965)
73. Wallach, D.P.: Studies on the GABA pathway. I.) The inhibition of gamma-aminobutyric acid-alpha-ketoglutaric acid transaminase in vitro and in vivo by U-7524 (amino-oxyacetic acid). Biochem. Pharmacol. **5**, 323 (1961)
74. Walters, J.R., Roth, R.H.: Effect of gamma-hydroxybutyrate on dopamine and dopamine metabolites in the rat striatum. Biochem. Pharmacol. **21**, 2111 (1972)
75. Yessaian, N.H., Arakelian, L.N.: The effect of gamma-aminobutyric acid on brain normetanephrine. J. Neurochem. **17**, 1689 (1970)

Effect of Acute Hyperammonemia on Metabolism and Cerebral Energy State[1]

G. BENZI[2], E. ARRIGONI, R.F. VILLA, and A. AGNOLI

Synopsis

The acute effect of hyperammonemia, kept at the level of 0.2 mM by adding ammonium acetate to circulating blood, was evaluated in the isolated dog brain in situ. The interference of the transmethylating system of S-adenosyl-L-methionine was also studied, by means of perfusions with S-adenosyl-L-methionine or adenosine (blood level 0.4 mM). The changes induced by the hyperammonemia syndrome on the *glutamate-ammonia system* (pyruvate, a-oxoglutarate, oxaloacetate, L-alanine, L-glutamate, L-aspartate, L-glutamine, NH_4^+) were evaluated. The cerebral detoxication of ammonia is connected with the formation of glutamine and, to a lesser extent, of alanine, balanced by a decrease in aspartate; glutamate, oxaloacetate, pyruvate, and a-oxoglutarate are unmodified or slightly modified. This shows that the glutamate-ammonia system is an integrated biochemical system where NH_4^+ represents the mobile element which modulates the interconversion of components.

Cerebral *glucidic metabolism* was largely activated by acute hyperammonemia, a marked increase in the free energy change being observed. A fraction of this energy not exceeding 10% can be ascribed to the synthesis of glutamine. Hyperammonemia induced a variation of the *resting transmembrane potential,* which becomes less negative. This event may explain why ammonia increases the excitability of nervous elements, to the point of inducing convulsions.

Introduction

A particularly interesting problem from an experimental and clinical standpoint is that of the specific ability of cerebral tissue to effect a detoxication of NH_4^+ during hyperammonemia syndromes. The glutamate-ammonia biochemical system is the most directly involved under these pathologic conditions; its participants are summarized in Figure 1.

Department of Pharmacology, University of Pavia (Italy) and Department of Neurology and Psychiatry, University of L'Aquila (Italy)

[1] A revised version of this paper has been accepted for publication in *Biochemical Pharmacology* (April 1977).

[2] Address for correspondence: Prof. Gianni Benzi, M.D., Istituto di Farmacologia, Facoltà di Scienze, Università di Pavia, Piazza Botta, 11; 27100 Pavia, Italy.

80

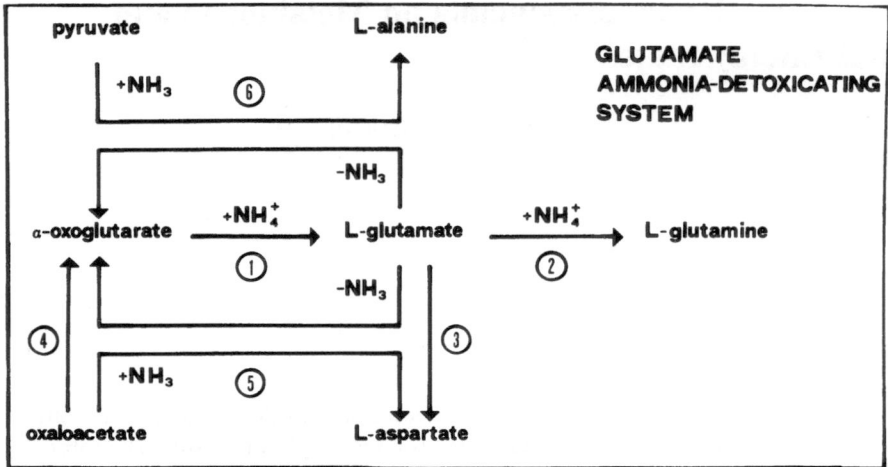

Fig. 1. Interconversion between the components of the glutamate-ammonia system

The basic event of this detoxication process consists of the conversion of a carboxyl acid into an amino acid, with ammonia uptake. a-Oxoglutarate thus undergoes an amination to L-glutamate, with uptake of an ammonia molecule; if L-glutamate undergoes a further amidation to L-glutamine, the uptake of another ammonia molecule takes place. In addition, Figure 1 shows that the glutamate-ammonia system is constituted by several carboxyl acids (a-oxoglutarate, pyruvate, oxaloacetate) and by several amino acids (L-glutamate, L-alanine, L-aspartate, L-glutamine) which can variously interconvert between themselves. Therefore, estimation of cerebral ammonia-detoxicating power must be carried out taking into account the integrated changes occurring in the various components, as a function of the amount of ammonia reaching the cerebral tissue. Thus the study of changes in a single participant yields nonsignificant data.

Another experimentally and clinically relevant remark is that some participants in the glutamate-ammonia system are intrinsically connected with the intermediate metabolism of glucides, as is summarized in Figure 2. Now, it is well known that at cerebral level glucidic metabolism is the main, if not the only, source of both the functional energy (related to functional activities) and the resting energy (related to the work of transport and concentration, as well as to biosynthetic work). These relationships between glycolysis (e.g., pyruvate) or Krebs cycle (e.g., a-oxoglutarate, oxaloacetate) intermediates and the glutamate-ammonia system require that the investigation be extended to the study of hyperammonemia effects also at the level of cerebral energy metabolism. It should further be noted that the function of cerebral cells requires a specific and uneven distribution of ions across the membranes. This ion distribution is improbable and each ion tends to randomize by passive diffusion along its electrochemical gradient. Consequently, energy is continuously required to maintain the steady state, and it is also important to evaluate the effect induced by hyperammonemia on the resting transmembrane potential. From a general point of view, it can be observed that, at cerebral level, in the hyperammonemia syndrome NH_4^+ ions can affect: (1) the specific ammonia-detoxicating power; (2) the energy state; (3) the resting transmembrane potential.

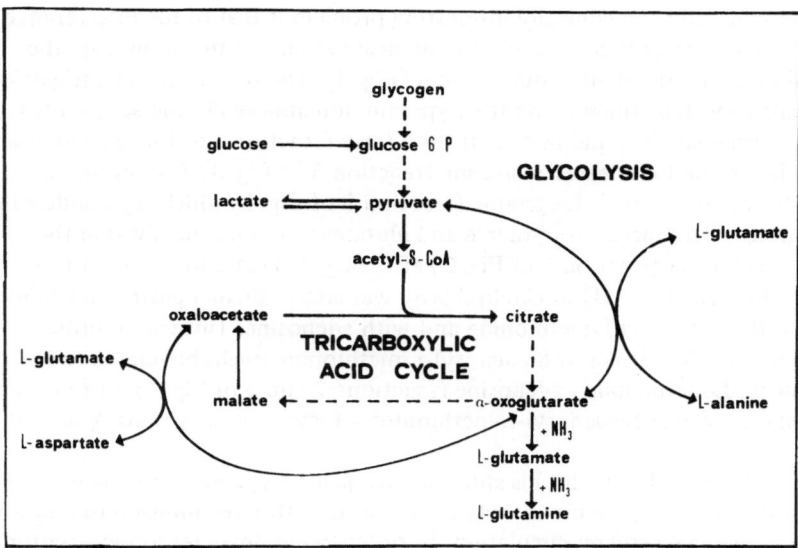

Fig. 2. Connection of some participants in the glutamate-ammonia system with glucidic metabolism

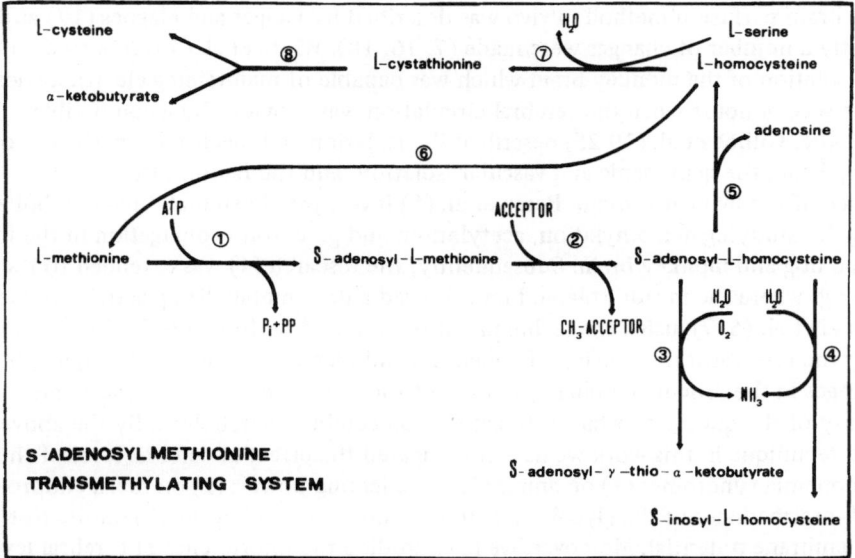

Fig. 3. The participants in the S-adenosyl-L-methionine transmethylating system

Another experimentally and clinically interesting problem is that of the interference of the glutamate-ammonia system with other biological systems of the brain, e.g., the S-adenosyl-L-methionine transmethylating system (Fig. 3). The reasons for investigating the effect of S-adenosyl-L-methionine on the hyperammonemia syndrome are that (a) this biological intermediate is found in an activated form (reaction 1 of Fig. 3); (b) as a cerebral metabolic product it releases adenosine (reaction 5 of Fig. 3); (c) during its metabolic transformation it yields L-cysteine (reaction 8 of Fig. 3) which, by condensing with a-oxoglutarate, yields mercaptopyruvate and glutamate; (d) it is involved in the transmethylation processes (reaction 2 of Fig. 3) affecting also catecholamines. In the present research, the effect of NH_4^+ at cerebral level was also evaluated during continuous perfusion both with S-adenosyl-L-methionine and with adenosine. This was in order to ascertain whether a possible effect of S-adenosyl-L-methionine might be related to the sequence S-adenosyl-L-methionine → adenosine (reactions 2 and 5 of Fig. 3), or to other metabolic sequences, such as S-adenosyl-L-methionine → L-cysteine (reactions 2, 5, 7, 8 of Fig. 3).

From a technical point of view, in this study in vivo acute hyperammonemia was induced by artificially increasing the blood NH_4^+ concentration through infusion of ammonium acetate to the regional cerebral circulation. In order to eliminate response variations due to the different blood levels of the various components, it is necessary that cerebral blood flow: (a) should be constant during each trial; (b) should have values $(ml.min^{-1}.g^{-1}$ of brain tissue) as close as possible during each trial. These conditions were met by the technique of in situ isolated brain perfusion because the composition of the perfusing blood, the blood flow rate, and blood oxygenation may be selected and maintained at will. A cat brain perfusion method in vivo was described by Geiger and Magnes (17) and subsequently a number of changes were made (7, 16, 18). White et al. (45) reported an anatomic isolation of the monkey brain which was capable of maintaining electrocortical activity for several hours when the cerebral circulation was provided by a compatible donor monkey. Gilboe et al. (19-25) described the isolation and mechanical perfusion of a viable dog head, the neurogenic and vascular isolation, and the factors affecting the maintenance of the living dog brain. Benzi et al. (5) investigated cerebral drug-metabolizing activity by studying demethylation, acetylation, and glucurono-conjugation in the in situ isolated dog and monkey brain. Subsequently, the research (4) was extended to the newborn dog, where the in situ isolated brain showed a drug-metabolizing activity related to age. Benzi et al. (6, 7), using the technique of the isolated perfused dog brain in situ, found that the intracarotid perfusion of ephedrine and nicergoline, but not lysergide, induced changes in the drug-metabolizing activity of the brain. This behavior was related to the ability of the quoted substances to act also on cerebral metabolism. By the above-mentioned technique in this work we have investigated the effect at cerebral level of the hyperammonemia syndrome: (1) on ammonia-detoxicating power; (2) on fuels, endproducts, and intermediates of the glycolytic pathway and citric acid cycle; (3) on the resting transmembrane potential. Moreover, we have studied the interference at cerebral level of the perfusion of S-adenosyl-L-methionine or adenosine on the effect induced by the hyperammonemia syndrome, evaluated by the three groups of parameters indicated above.

Material and Methods

Animals and Anesthesia. The experiments were carried out on female beagle dogs aged 240-360 days and weighing from 12.5 to 15.6 kg. Before the experiments, the dogs were maintained under the same environmental conditions (22 ± 1º; relative humidity: 60 ± 5%) and were fed only a standard diet with water ad libitum. The surgical procedure was performed on animals preanesthetized with urethane (0.4 g/kg i.p.). EEG pattern was used to determine the degree of anesthesia (15), which was induced and maintained only during the surgical procedure by chloralose (20-40 mg/kg i.v.). The anesthetics affect labile phosphates and the extra- and intracellular brain lactate and pyruvate concentrations (31, 39). Therefore the restoration of normal EEG pattern was used to indicate the removal of the anesthetic from the brain before the start of the experiment on cerebral energy metabolism. The dogs were artificially ventilated with an intratracheal Warne tube. During the operative procedure and the experiment, the animals were immobilized by intravenous injection of gallamine triethiodide (2-3 mg/kg).

Operative Procedure. The operative procedure (6) consisted mainly in the isolation of the external jugular veins and the common carotid arteries, with ligature of all their branches (except the internal carotid arteries) and the vertebral vessels. The numerous muscular branches arising from the vertebral vessels, the anastomosis between vertebral and carotid arteries, the anastomosis between vertebral and jugular veins, the internal jugular veins, the vascular branches of the neck, the vessels running under the carotid arteries and vagus nerves, and the zygomatic, maxillary, auricular, and supraorbital vessels were all occluded by ligature or compression. The occlusion of the sinus columnae vertebralis was made by opening the rachis in C2 and compressing the venous vessels around the spinal cord.

Both of the isolated jugular veins were ligated, cannulated, and connected to the venous reservoir of the pump-oxygenator system (through the gravitational flow). Both of the isolated carotid arteries were also cannulated and connected to the pump-oxygenator system. Systemic arterial blood pressure was measured from a cannula inserted into a femoral artery. The head was fixed to a head holder with the confluence of the cerebral venous sinuses approximately 10 cm higher than the heart. Monopolar electrodes were inserted in the left and right frontal, parietal, and occipital areas. The electroencephalogram as well as both systemic and cerebral perfusion pressure were recorded on a 12-channel polygraph (Physioscript EE12-Schwarzer). After a longitudinal incision, a 2.5 cm diameter hole was made in the frontoparietal area. A plastic funnel was fitted into the hole and the skin was tightly sutured around the funnel; subsequently, the plastic funnel was sealed with a rubber stopper and thermally insulated.

Brain Pump-Oxygenator System. The brain perfusion apparatus employed consisted of a venous reservoir, an oxygenator with a gasmeter, a roller-type pump with a flowmeter, two blood filters (polyester staple 1622 Montefibre), an apparatus to eliminate blood foam, a perfusion pressure regulator with a manometer, and a blood exchanger with a telethermometer. Before the extracorporeal perfusion, the pump-oxygenator system was filled with 500 ml of heparinized compatible blood. The blood was obtained 20 min prior to use prevent the accumulation of lactic acid, which would subsequently require a considerable degree of neutralization.

Before the perfusion, the blood was filtered through polyester staple and adjusted to pH 7.35 using 1 M sodium bicarbonate. A flow of a O_2 + CO_2 mixture (95:5), maintained at the rate of 5 l.min^{-1}, was passed into the oxygenator during the extracorporeal brain perfusion. The blood flow rate was kept between 41 and 43 ml.min^{-1}, the pressure being equal to the initial systemic pressure of the animal.

Analytical Techniques. The arteriovenous differences of metabolites across the brain, glucose, glutamate, glutamine, alanine, O_2 and NH_4^+ uptake, and lactic acid formation were calculated from simultaneously drawn arterial and venous blood samples. In fact (25), the rate of lactic acid formation by the erythrocyte system is sufficiently high to cause interference with the precise determination of the lactic acid released by the brain. Such interference is substantially decreased by drawing arterial and venous samples simultaneously and precipitating the blood protein immediately. By continuous infusion of glucose (25 μmol.min^{-1}) to the venous reservoir, the glucose concentration in the blood was maintained at a constant level.

To evaluate the metabolites in the brain, at the set time, the motor area of the cortex was frozen in situ by pouring liquid nitrogen into the plastic funnel fitted into the cranial vault. The cortical portion of the frozen brain was cut and removed using a rotating cold hollow tube during continuous irrigation with liquid nitrogen. The frozen cerebral material was then immersed into liquid nitrogen for 10-15 min and quickly (3-4) powdered by a precooled automatic apparatus (Microdismembrator-Braun) using frozen 1.23 M perchloric acid. The subsequent steps of analytical procedure were carried out in a cooled box at 0°-5°C until a perchlorate-free extract was obtained.

Metabolites were determined by enzymatic techniques: alanine (26); NH_4^+ (32); aspartate (9); citrate (37); glycogen (34); glucose (10); glutamate (12); glutamine (35); lactate (28); malate (36); a-oxoglutarate (8); O_2 (42); pyruvate (14). Where possible, several metabolites were determined in the same cuvette by sequential addition of the appropriate enzymes. The oxaloacetate concentration was calculated from the formula:

$$[\text{oxaloacetate}] = \frac{[\text{pyruvate}] \times [\text{malate}]}{[\text{lactate}]} \times \frac{K_{\text{MDH}}}{K_{\text{LDH}}}$$

where K_{LDH} and K_{MDH} are the equilibrium constants of *lactate dehyldrogenase* and *malate dehydrogenase* respectively (46).

The hyperammonemia syndrome was induced by infusion of ammonium acetate into the extracorporal circuit at an initial concentration of 2×10^{-4} M. The level of NH_4^+ in the perfusion arterial blood was kept constant by the continuous addition of 6.5 μmol.min^{-1}, during the 20 min of the syndrome. S-adenosyl-L-methionine (Bioresearch, Milano) or adenosine (Hoechst, Frankfurt) were present in the perfusion circuit at an initial concentration of 4×10^{-4} M; during the 40 min of the experiment, they were continuously infused (1 μmol.min^{-1}).

Our investigation was focused on three series of biological events taking place in brain tissue:

1. the NH_4^+-detoxicating power of the glutamate-ammonia system
2. the brain energy balance related to the intermediate metabolism of glucides

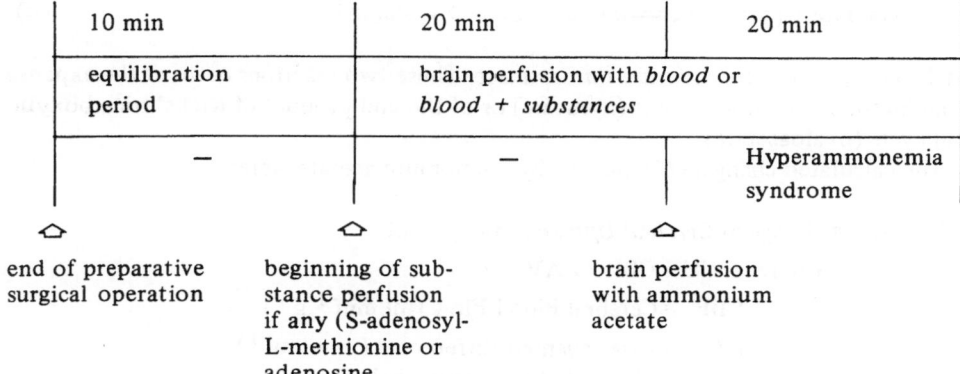

10 min	20 min	20 min
equilibration period	brain perfusion with *blood* or *blood + substances*	
−	−	Hyperammonemia syndrome

end of preparative surgical operation	beginning of sub-stance perfusion if any (S-adenosyl-L-methionine or adenosine	brain perfusion with ammonium acetate

Fig. 4. *Scheme of the Experiment*

3. the resting transmembrane potential, indirectly obtained by applying the Nernst equation.

In order to define the effect induced by the hyperammonemia syndrome, control values were taken as base values, changes (Δ') in the parameters examined being calculated by difference.

Evaluation of the NH_4^+-Detoxicating Power of the Glutamate-Ammonia System

This evaluation was calculated by the balance of changes (Δ') in the arterovenous levels and/or in the cerebral concentrations of the participants in the glutamate ammonia detoxicating system shown in Figure 1. Some participants of this system do not have any ammonia radicals (pyruvate, α-oxoglutarate, oxaloacetate), while others have one (L-glutamate, L-alanine, L-aspartate) or two (L-glutamine).

According to reaction 1, α-oxoglutarate may be converted to glutamate by amination:

$$\alpha\text{-oxoglutarate} + NADH + H^+ + NH_3 \longrightarrow L\text{-glutamate} + NAD^+ + H_2O \tag{1}$$

catalyzed by *glutamate dehydrogenase.* This reaction is practically irreversible, since the presence in the brain of any detectable amount of NH_4^+ would oppose the reverse reaction. According to reaction 2, glutamate may be further converted to glutamine by amidation:

$$L\text{-glutamate} + ATP + NH_3 \xrightarrow{Mg^{2+}} L\text{-glutamine} + ADP + P_i \tag{2}$$

catalyzed by *glutamine synthase.* This reaction requires the utilization of ATP. According to reactions 3 and 4, L-glutamate may be converted to α-oxoglutarate according to freely reversible transamination reactions:

$$L\text{-glutamate} + oxaloacetate \rightleftharpoons \alpha\text{-oxoglutarate} + L\text{-aspartate} \tag{3}$$

catalyzed by *glutamate-oxaloacetate transaminase,* and

$$L\text{-glutamate} + pyruvate \rightleftharpoons a\text{-oxoglutarate} + L\text{-alanine} \tag{4}$$

catalyzed by *glutamate-pyruvate transaminase*. These two reactions occur at the expense of an intermediate of glycolysis (pyruvate) or of the end product of Krebs' tricarboxylic acid cycle (oxaloacetate).

The calculated changes (Δ') induced by ammonium acetate were:

◊ $\Delta'CU$ = change in *Cerebral Uptake* ($nmol.g^{-1}.min^{-1}$)

where: $\Delta'CU = CBF \times \Delta AV$

CBF = Cerebral Blood Flow ($ml.min^{-1}.g^{-1}$)

ΔAV = Arteriovenous differences ($nmol.ml^{-1}$) of metabolites across the brain.

◊ $\Delta'CTC$ = change in *Cerebral Tissular Concentration* ($nmol.g^{-1}.min^{-1}$)

$\Delta'CTC = (CTC_{amm} - CTC_{cont}) \cdot t^{-1}$

where: CTC_{amm} = cerebral tissular concentrations ($nmol.g^{-1}$) of metabolites after ammonium acetate perfusion;

CTC_{cont} = cerebral tissular concentrations ($nmol.g^{-1}$) of metabolites in control conditions;

t = time (min).

◊ $\Delta'CMR$ = change in *Cerebral Metabolic Rate* ($nmol.g^{-1}.min^{-1}$)

$\Delta'CMR = \Delta'CU - \Delta'CTC$

◊ $\Delta'CMZR_{NH_4^+}$ = change in *Cerebral MetaboliZing Rate for* NH_4^+ ($nmol.g^{-1}.min^{-1}$)

$\Delta'CMZR_{NH_4^+} = \Delta'CTC_{glutamate} + \Delta'CTC_{aspartate} + \Delta'CTC_{alanine} + 2\Delta'CTC_{glutamine}$

◊ $MPI_{NH_4^+}$ = *Metabolizing Power Index* for NH_4^+

$$MPI_{NH_4^+} = \frac{\Delta'CMZR_{NH_4^+}}{\Delta'CMR_{NH_4^+}}$$

This evaluation is particularly important since it indicates the amount of detoxicated NH_4^+ by the intervention of the glutamate-ammonia system with regard to the NH_4^+ afferent to the cerebral tissue.

Computation of Cerebral Energy Change Caused by NH_4^+

This evaluation was obtained from the balance of variations in the arteriovenous and/or cerebral concentrations of substrates and of intermediates related to cerebral glucidic metabolism.

The calculated changes (Δ') induced by ammonium acetate were:

◊ Δ'CU = change in *Cerebral Uptake* (nmol.g^{-1}.min^{-1})

$$\Delta'CU = CBF \ (ml.min^{-1}.g^{-1}) \ \times \ \Delta AV \ (nmol.ml^{-1})$$

◊ Δ'CTC = change in *Cerebral Tissular Concentration* (nmol.g^{-1}.min^{-1})

$$\Delta'CTC = [CTC_{amm}(nmol.g^{-1}) - CTC_{cont}(nmol.g^{-1})] \cdot t^{-1}(min)$$

◊ Δ'CMR = change in *Cerebral Metabolic Rate* (nmol.g^{-1}.min^{-1})

$$\Delta'CMR = \Delta'CU - \Delta'CTC$$

(For the details of the above mentioned changes we refer to what has been said in the previous paragraph with regard to ammonium-detoxicating power)

◊ Δ'CAnMR$_{glucose}$ = change in *Cerebral Anaerobic Metabolic Rate for the glucose units* (nmol.g^{-1}.min^{-1})

Δ'CAnMR$_{glucose}$ = -0.5 Δ'CMR$_{lactate}$

where:

Δ'CMR$_{lactate}$ = change in *Cerebral Metabolic Rate for lactate*. In fact each mole of glucose (or of glucosyl units derived from glycogen) anaerobically yields two moles of lactate.

◊ Δ'CAeMR$_{glucose}$ = change in *Cerebral Aerobic Metabolic Rate for glucose* (nmol.g^{-1}.min^{-1})

Δ'CAeMR$_{glucose}$ = Δ'CMR$_{glucose}$ + Δ'CMR$_{glycogen}$ - Δ'CAnMR$_{glucose}$

◊ Δ'[ΔG°] = change in the *cerebral free energy* (mcal.g^{-1}.min^{-1})

For the measurement of the energy change induced by NH_4^+ in the brain system, we used the function ΔG or *Gibbs free energy*, defined as:

$$\Delta G = \Delta E + P \Delta V - T \Delta S \tag{5}$$

where: P and V are the pressure on, and the volume of, the system, respectively; ΔE represents the changes in internal energy of the system during the biochemical process; T is the absolute temperature; ΔS represents the changes in entropy for the system. It should also be noticed that for the chemical reactions implied in the present experimental model, the PΔV term makes a negligible conribution to energetics because the volume change, ΔV, during transformation is very small. Consequently, Equation (5) may be replaced by

$$\Delta G = \Delta E - T \Delta S \tag{6}$$

To simplify further actual numerical computations, we used the *standard* free energy change (ΔG°), assuming the value of -47 kcal.mol^{-1} for the reduction of glucose to lactate and the value of -686 kcal.mol^{-1} for the oxidation of glucose to H_2O and CO_2. Obviously, the absolute ΔG value for reaction occurring at a physiologic concentration may be different from the value of the change in free energy for standard conditions. Consequently, the computated values of free energy change in our experimental model may be regarded as *relative* and utilized only for the evaluation of the changes (Δ') occurring between the control condition and the NH_4^+-induced condition: $\Delta'G^{\circ'}$). This procedure is incapable of determining the absolute value of each experimental condition. In practice, however, this limitation is not really a handicap, since interest is generally focused on the bioenergetic change induced by NH_4^+, and the control state may be chosen as the convenient reference point.

Computation of the Resting Transmembrane Potential

For the indirect calculation of the resting transmembrane potential (E_h) we used the Nernst equation:

$$E_h = -\frac{RT}{F} \ln \left(\frac{[K^+]_{int} + a[NH_4^+]_{int}}{[K^+]_{ext} + a[NH_4^+]_{ext}}\right) \text{ mV}$$

where: R = gas law constant (1.987 cal.mol^{-1}.deg^{-1})
  T = absolute temperature
  F = Faraday constant (23,061 cal.vol^{-1}.equiv^{-1}).
  a = 0.2

The subscripts *int* and *ext* refer to the intracellular and extracellular concentrations respectively, and a is the permeability of NH_4^+ relative to K^+. We used the blood $[NH_4^+]$ and $[K^+]$ as an approximation of those existing in brain extracellular fluid.

Results

Ammonium acetate, added to the cerebral perfusing blood (0.2 mM) causes a derangement in the components of the cerebral *glutamate-ammonia system,* consisting (Table 1) of a significant increase in glutamine and alanine, and a significant decrease of both aspartate and oxaloacetate, while glutamate, pyruvate and a-oxoglutarate remain unchanged. A greater interest lies in computations shown in Table 2, where changes in cerebral uptake, cerebral tissular concentration, cerebral metabolic rate, and cerebral metabolizing rate for ammonia are calculated for the various metabolites. It is thus possible to observe that ammonia detoxication is brought about by glutamine and, to a lesser degree, by alanine. Concurrently, the detoxicating role of glutamate and aspartate becomes less important. This behavior is probably due to the amidation of glutamate to glutamine. In addition, Table 2 also shows that adenosine perfusion (0.4 mM) does not remarkably modify the ammonium-detoxicating power of the brain tissue, whereas the perfusion with S-adenosyl-L-methionine (0.4 mM) increases the uptake and metabolization of NH_4^+, even though no changes in the cerebral ammonia concentration take place. The most significant result of the pretreatment perfusion with S-adenosyl-L-methinine is an increase in the ammonium-

Table 1. Action of ammonium acetate (added to the cerebral perfusing blood: 2×10^{-4} M) on some components of the *glutamate-ammonia system* evaluated in cerebral tissue and by arteriovenous differences across the brain. The values are means (± standard errors) of six preparations of beagle dog. The symbol (●) indicates statistical significance as compared with control conditions: (*b*), (*c*) and (*d*) versus (*a*). The symbol (▲) in the hyperammonemic animals indicates statistical significance as compared with dogs not infused by S-adenosyl-L-methionine or adenosine: (*c*) and (*d*) versus (*b*). Level of statistical significance: $P < 0.05$

Parameters		Unit of measurement	Control conditions (*a*)	Perfusion with ammonium acetate after and during the infusion with (*b*)	S-adenosyl-L-methionine $(4 \times 10^{-4}\,M)$ (*c*)	Adenosine $(4 \times 10^{-4}\,M)$ (*d*)
From arteriovenous differences across the brain	Cerebral blood flow	ml.min^{-1}.g^{-1}	0.494 ± 0.011 ●	0.492 ± 0.010 ●	0.477 ± 0.012	0.485 ± 0.015
	NH$_4^+$ uptake	nmol.min^{-1}.g^{-1}	2 ± 1 ●	78 ± 2 ●	94 ± 3 ●▲	81 ± 3 ●
	Glutamate uptake		<1	<1	<1	<1
	Glutamine uptake		1.2 ± 0.3	1.6 ± 0.4	1.1 ± 0.3	<1
	Alaine uptake		1.1 ± 0.6	<1	1.5 ± 0.4	<1
Concentration in brain	NH$_4^+$	μmol.g^{-1}	0.23 ± 0.04	0.42 ± 0.01 ●	0.44 ± 0.02 ●	0.39 ± 0.03 ●
	Glutamate		10.20 ± 0.41	10.09 ± 0.32	10.12 ± 0.11	10.06 ± 0.06
	Glutamine		5.43 ± 0.08	6.29 ± 0.02 ●	6.51 ± 0.12 ●	6.39 ± 0.07 ●
	Alanine		0.435 ± 0.011	0.585 ± 0.011 ●	0.595 ± 0.007 ●	0.635 ± 0.017 ●▲
	α-Oxoglutarate		0.199 ± 0.013	0.192 ± 0.006	0.202 ± 0.003	0.195 ± 0.002
	Asparate		2.44 ± 0.10	1.93 ± 0.04 ●	1.80 ± 0.06 ●	1.80 ± 0.10 ●
	Pyruvate		0.100 ± 0.008	0.104 ± 0.004	0.102 ± 0.003	0.103 ± 0.002
	Oxaloacetate	nmol.g^{-1}	3.3 ± 0.2	2.7 ± 0.1 ●	2.5 ± 0.1 ●	2.8 ± 0.1 ●

Table 2. Changes (Δ') induced by ammonium acetate (added to cerebral perfusing blood: 2×10^{-4} M) in the *glutamate-ammonia system* of the brain as compared with control conditions. The calculated parameters are:

Δ'CU = change in Cerebral Uptake

Δ'CTC = change in Cerebral Tissular Concentration

Δ'CMR = Δ'CU - Δ'CTC = change in Cerebral Metabolic Rate

Δ'CMZR$_{NH_4^+}$ = change in Cerebral Metabolizing Rate for NH$_4^+$

The symbol ($\bullet$) indicates the statistical significance ($P < 0.05$) as compared with perfusion without S-adenosyl-L-methionine or adenosine: *(b)* and *(c)* versus *(a)*

Cerebral infusion with	Metabolite	Δ'CU	Δ'CTC	Δ'CMR	Δ'CMZR$_{NH_4^+}$	Metabolizing Power Index for NH$_4^+$
			nmol.g^{-1}.min^{-1}			
	NH$_4^+$	+ 76	+ 9.5	+ 66.5		
	Glutamate	< 0.5	− 5.5		− 5.5	
	Glutamine	< 0.5	+ 43.0		+ 86.0	
	Alanine	< 0.5	+ 7.5		+ 7.5	
	Aspartate		− 25.5		− 25.5	
	Pyruvate		< 0.5			
	α-Oxoglutarate		< 0.5			
	Oxaloacetate		< 0.5			
(a)	Total				+ 62.5	0.94
	NH$_4^+$	+ 92$\bullet$	+ 10.5	+ 81.5$\bullet$		
	Glutamate	< 0.5	− 4		− 4	
	Glutamine	< 0.5	+ 54$\bullet$		+108$\bullet$	
S-adenosyl-L-methione	Alanine	< 0.5	+ 8		+ 8	
	Aspartate		− 32$\bullet$		+ 32 $\bullet$	
	Pyruvate		< 0.5			
	α-Oxoglutarate		< 0.5			
	Oxaloacetate		< 0.5			
(b)	Total				+ 80 $\bullet$	0.98
	NH$_4^+$	+ 79	+ 8	+ 71		
	Glutamate	< 0.5	− 7		− 7	
	Glutamine	< 0.5	+ 48		+ 96	
	Alanine	< 0.5	+ 10		+ 10	
Adenosine	Aspartate		− 32 $\bullet$		− 32 $\bullet$	
	Pyruvate		< 0.5			
	α-Oxoglutarate		< 0.5			
	Oxaloacetate		< 0.5			
(c)	Total				+ 67	0.94

Table 3. Action of ammonium acetate (added to the cerebral perfusing blood: 2×10^{-4} M) on some parameters related to glucidic metabolism and evaluated in cerebral tissue and by arterovenous differences across the brain. The values are means ($\pm$ standard errors) of six preparations. The symbol ($\bullet$) indicates statistical significance as compared with control conditions. (b), (c) and (d) versus (a). The symbol ($\blacktriangle$) in the hyperammonemic animals indicates statistical significance as compared with dogs not infused by S-adenosyl-L-methionine or adenosine: (c) and (d) versus (b). Level of statistical significance: $P < 0.05$

Parameters		Unit of measurement	Control condition (a)	Perfusion with ammonium acetate after and during the infusion with — (b)	S-adenosyl-L-methionine (4×10^{-4} M) (c)	Adenosine (4×10^{-4} M) (d)
From arterovenous differences across the brain	Cerebral blood flow	ml.min^{-1}.g^{-1}	0.494 ± 0.011	0.492 ± 0.010	0.477 ± 0.012	0.485 ± 0.015
	Glucose uptake	μmol.g^{-1}.min^{-1}	0.285 ± 0.008	0.322 ± 0.006 ●	0.319 ± 0.014 ●	0.325 ± 0.007 ●
	Lactate release		0.038 ± 0.002	0.075 ± 0.005 ●	0.074 ± 0.005 ●	0.081 ± 0.005 ●
	Oxygen uptake		1.64 ± 0.07	1.73 ± 0.04	1.76 ± 0.02	1.75 ± 0.03
Concentration in brain	Glycogen	μmol.g^{-1}	3.48 ± 0.16	2.12 ± 0.03 ●	2.44 ± 0.07 ●▲	2.34 ± 0.06 ●
	Glucose		1.44 ± 0.15	2.66 ± 0.07	2.28 ± 0.06 ●▲	2.48 ± 0.06 ●
	Pyruvate	μmol.g^{-1}	0.100 ± 0.008	0.104 ± 0.004	0.102 ± 0.003	0.103 ± 0.002
	Lactate		1.64 ± 0.15	2.23 ± 0.07 ●	1.96 ± 0.04 ●	2.06 ± 0.04 ●
	ATP	μmol.g^{-1}	2.21 ± 0.04	2.26 ± 0.05	2.16 ± 0.07	2.20 ± 0.07
	ADP		0.48 ± 0.02	0.45 ± 0.01	0.52 ± 0.03	0.43 ± 0.02
	AMP		0.06 ± 0.002	0.09 ± 0.003	0.05 ± 0.003	0.08 ± 0.002
	Creatine phosphate		4.82 ± 0.12	4.40 ± 0.36	4.76 ± 0.20	4.59 ± 0.23

Table 4. Changes (Δ') induced by ammonium acetate (added to the cerebral perfusing blood: 2×10^{-4} M) in some parameters related to the glucidic metabolism of the brain as compared with control condition. The calculated parameters (see also *Material and Methods*) are:

$\Delta'CU$ = change in Cerebral Uptake;

$\Delta'CTC$ = change in Cerebral Tissular Concentration;

$\Delta'CMR = \Delta'CU - \Delta'CTC$ = change in Cerebral Metabolic Rate;

$\Delta'CAnMR_{glucose} = -5 \, (\Delta'CMR_{lactate})$ = change in Cerebral Anaerobic Metabolic Rate for glucose;

$\Delta'CAeMR_{glucose} = \Delta'CMR_{glucose} + \Delta'CMR_{glycogen} - \Delta'CAnMR_{glucose}$ = change in Cerebral Aerobic Metabolic Rate for clugose;

$\Delta'(\Delta G^{\circ})$ = change in Gibbs free energy

The symbol (●) indicates statistical significance as compared with perfusion without S-adenosyl-L-methionine or adenosine: (b) and (c) versus (a). Level of statistical significance: $P < 0.05$

Cerebral infusion with	Metabolite	$\Delta'CU$	$\Delta'CTC$	$\Delta'CMR$	$\Delta'CAnMR_{glucose}$	$\Delta'CAeMR_{glucose}$	$\Delta'(\Delta G^{\circ})$	
				nmol.g^{-1}.min^{-1}			mcal.g^{-1}.min^{-1}	%
—	glucose	+ 37	+ 61	− 24			− 7.55	83
	glycogen		− 68	+ 68		+ 11	− 1.55	17
	lactate	− 37	+ 29	− 66	+ 33			
(a)	total						− 9.10	100
S-adenosyl-L-methionine (4×10^{-4} M)	glucose	+ 34	+ 42 ●	− 8 ●		+ 18 ●	− 12.35 ●	91 ●
	glycogen		− 52 ●	+ 52 ●			− 1.22 ●	9 ●
	lactate	− 36	+ 16 ●	− 52 ●	+ 26 ●			
(b)	total						− 13.57 ●	100
Adenosine (4×10^{-4} M)	glucose	+ 40	+ 52	− 12 ●		+ 13	− 8.92	86
	glycogen		− 57	+ 57			− 1.50	14
	lacate	− 43	+ 21	− 64	+ 32			
(c)	total						− 10.42	100

Table 5. Action of ammonium acetate (added to cerebral perfusing blood: 2×10^{-4} M) on some parameters related to the *resting transmembrane potential* and evaluated both in arterial blood and in cerebral tissue. The values are means (± standard errors) of six preparations of beagle dog. The symbol (●) indicates statistical significance as compared with control conditions. The symbol (▲) in hyperammonemic dogs indicates statistical significance as compared with animal not infused by S-adenosyl-L-methionine or adenosine: *(c)* and *(a)* versus *(b)*. Level of statistical significance: $P < 0.05$

Parameters	Unit of measurement	Control conditions	Perfusion with ammonium acetate after and during the infusion with		
			— *(a)*	S-adenosyl-L-methionine $(4 \times 10^{-4}$ M) *(b)*	Adenosine $(4 \times 10^{-4}$ M) *(c)*
K^+ arterial		4.12 ± 0.16	4.71 ± 0.03 ●	4.88 ± 0.04 ●	4.77 ± 0.03 ●
K^+ cerebral		120.1 ± 1.5	116.5 ± 0.5	114.2 ± 0.4	116.4 ± 0.4
NH_4^+ arterial	$\mu mol \cdot g^{-1}$	0.022 ± 0.003	0.214 ± 0.007 ●	0.226 ± 0.004 ●	0.221 ± 0.004 ●
NH_4^+ cerebral		0.23 ± 0.04	0.42 ± 0.01 ●	0.44 ± 0.02 ●	0.39 ± 0.03 ●
Calculated resting transmembrane potential	mV	-90.88 ± 1.26	-86.20 ± 0.63 ●	-84.63 ± 0.28 ●	-85.79 ± 0.17 ●

The resting transmembrane potential was calculated as: $-\dfrac{RT}{F} \ln \left(\dfrac{[K^+]_{int} + a[NH_4^+]_{int}}{[K^+]_{ext} + a[NH_4^+]_{ext}} \right)$, where the subscripts *int* and *ext* refer to intracellular (cerebral) and extracellular (arterial) concentrations respectively, and a is the permeability of NH_4^+ relative to K^+.

detoxicating power of glutamine, matched by a decrease in that of aspartate. Under the three experimental conditions listed in Table 2, it can be seen that the ratio between ammonium-detoxicating power and ammonia cerebral metabolic rate, i.e., the Metabolizing Power Index for NH_4^+, is always very close to 1. This index, in fact, has a value of 0.94 in hyperammonemia, and values of 0.98 or 0.94 in the hyperammonemia developing during perfusion with S-adenosyl-L-methionine and adenosine, respectively.

The effect of ammonium acetate on cerebral intermediate *metabolism of glucides* is summarized in Table 3, which shows a significant increase both in glucose uptake and in the production of lactate. At cerebral level, an increase in the concentration of glucose and lactate can be observed, while glycogen depletion occurs and tissue levels of pyruvate remain unchanged. For the purpose of calculating the changes in free energy induced by the various experimental conditions, the data relative to the balance of glucose, glycogen, and lactate are reported in Table 4. From these computations it can be seen that hyperammonemia induces a remarkable increase in the cerebral free energy. The intervention of S-adenosyl-L-methionine increases the free energy change, not so much by causing an increase in the substrate uptake as by displacing the increased glucidic metabolism towards aerobic degradation.

From the values of the arterial and cerebral concentrations of K^+ and NH_4^+ (see Table 5) the *resting transmembrane potential* was indirectly calculated. Hyperammonemia makes this potential less negative; the extent of this effect is slightly increased by the cerebral perfusion with S-adenosyl-L-methionine and, to a lesser degree, with adenosine.

Discussion

This research performed on the motor area of the beagle dog brain cortex shows that arterial NH_4^+ increase produces various cerebral metabolic derangements. As concerns the glutamate-ammonia system (Fig. 1), the increase in glutamine synthesis represents the major mechanism in ammonia detoxication, as has been proposed also on the basis of previous studies. In fact, in cat brain (11) ^{15}N from ^{15}N-ammonium acetate was found to be preferentially incorporated into the amino group of glutamine, as compared with the amino group of glutamate. Furthermore, in the presence of an increased level of ammonia, in the brain cortex of the cat (44) the specific activity of tissue $^{14}CO_2$ for glutamine increased, without a corresponding decrease in specific activities for glutamate or aspartate. In agreement with the results obtained in immediately frozen (43) rat brain (29), in our study of NH_4^+-induced derangement of the glutamate-ammonia system in the motor area of dog brain cortex we found (Table 2) that: (1) the cerebral metabolizing rate for NH_4^+ was increased; (2) much NH_4^+ was incorporated into glutamine and a lesser amount into alanine: in fact the NH_4^+-induced synthesis of extra glutamine was about 43 nmol.g^{-1}.min^{-1}, while the synthesis of extra alanine was about 7.5 nmol.g^{-1}.min^{-1}; (3) the NH_4^+-induced change in the cerebral metabolic rate of glutamate, oxaloacetate, pyruvate, and a-oxoglutarate was negligible; (4) the cerebral metabolic rate of aspartate was decreased; (5) pretreatment with S-adenosyl-L-methionine produced an increase in the NH_4^+-induced synthesis of extra glutamine, consistent with the lowered metabolic rate of aspartate. Such results confirm that the glutamate-ammonia system forms an integral whole (3) in which fluctuations in the value of one member, such as ammonia, are compensated for by corresponding fluctuations in the values of other members. At any rate, the glutamate-ammonia system plays a modulatory role in the cerebral process of NH_4^+ detoxication. Indeed, the metabolizing power index for NH_4^+ showed a value of 0.94 during the hyperammonemia syndrome, and values of 0.98 or 0.94 during the hyperammonemia syndrome elicited in the course of the brain perfusion with S-adenosyl-L-methionine or adenosine.

The cerebral glutamate-ammonia system is segregated into *small* and *large* compartments. The *large* compartment appears to contain the major pools of glutamate and aspartate, and to lie in the neurons; the *small* compartment is associated with the major pools and sites of biosynthesis of glutamine and lies in the glia (2). On the other hand, the data obtained by density-gradient-centrifugation methods suggest the existence of at least two subpopulations of mitochondria, one characterized by the presence of a relatively large amount of glutamate dehydrogenase, and the other by a relatively large amount of citric acid-cycle enzymes (38, 41). Ammonia is incorporated into an amino group by the activity of glutamate dehydrogenase, which has a restricted localization in the brain (40), in one or more small glutamate compartments. In the present research, pretreatment with S-adenosyl-L-methionine gave further evidence of the homeostatic mechanism regulating the NH_4^+-induced interconversion of the members of the glutamate-ammonia system. This supports the suggestion (3) that there exists a cycle of metabolic events linking the glia and neurons, the glutamate → glutamine interconversion being a major linking process. Consequently it is possible that in physiologic and pathologic situations the ammonium ion plays a basic role in the linkage of metabolic processes between cerebral compartments.

The interference of S-adenosyl-L-methionine on NH_4^+ detoxication is probably related to the formation of L-cysteine and its subsequent condensation with a-oxoglutarate to form mercaptopyruvate and glutamate, which may then be further converted to glutamine. In fact (Fig. 2), S-adenosyl-L-methionine provides the methyl groups for the acceptors (13) with the formation of S-adenosyl-L-homocysteine, which is split to give L-homocysteine and adenosine; L-homocysteine may then undergo remethylation to L-methionine or enter the transsulfuration pathway by the formation of L-cysteine (27, 33). The interference of S-adenosyl-L-methionine in NH_4^+ detoxication is probably unrelated to adenosine production, because of the inactivity of the pretreatment with adenosine itself, as indicated in Table 2.

As for the problem of the change induced by NH_4^+ in *cerebral energetics*, the view that NH_4^+ interferes with "high-energy bond" phosphate seems unlikely, as some biochemical researches have shown no change in cerebral energy charge potential of the adenylate pool (29, 30). In the present research (Table 3), adenine nucleotides concentration was unchanged, indicating no decrease in the phosphorylation state of the brain during acute hyperammonemia. Furthermore, we have observed that the hyperammonemia syndrome induces an increase in the change Gibbs free energy (Table 4). Therefore, the current opinion according to which acute hyperammonemia would induce a drop of the energy storage at the cerebral level does not appear to be well-founded. The excess of free energy induced by the increased rate of glucidic metabolism can be utilized by a process which occurred as a consequence of increased NH_4^+, such as: incorporation of NH_4^+ into glutamine, increase of amino acids transport to the brain, or ionic effects of ammonia.

As for NH_4^+ incorporation into glutamine, let us consider this amidation reaction, catalyzed by *glutamine synthase:*

$$\text{L-glutamate} + NH_3 \longrightarrow \text{L-glutamine} \tag{7}$$

where it is known that at 37°C and for pH = 7

$$K' = \frac{[\text{L-glutamine}]}{[\text{L-glutamate}][NH_3]} = 4.45 \times 10^{-3} \tag{8}$$

hence: $\Delta G^{\circ\prime} = - RT \ln K' = 3.33 \text{ kcal.mol}^{-1}$

Since $\Delta G > 0$, reaction (7) cannot occur spontaneously. Its occurrence is, however, made possible by the thermodynamic intervention of ATP:

$$\text{MgATP} + \text{L-glutamate} + NH_3 \longrightarrow \text{MgADP} + P_i + \text{L-glutamine} \tag{9}$$

where it is known that at 37°C and for pH = 7

$$K' = \frac{[\text{MgADP}][P_i][\text{L-glutamine}]}{[\text{MgATP}][\text{L-glutamate}][NH_3]} = 4 \times 10^2 \tag{10}$$

hence: $\Delta G^{\circ\prime} = - RT \ln K' = -3.69 \text{ kcal.mol}^{-1}$

Since $\Delta G < 0$, reaction (9) takes place spontaneously from left to right. In this case the energy expenditure is 7 kcal.mol^{-1}, since, by substituting the value of (8) in (10), one obtains:

$$K' = \frac{[\text{MgADP}][P_i]}{[\text{MgATP}]} = 8.99 \times 10^4 \tag{11}$$

hence: $\Delta G^{\circ\prime} = - 7.03 \text{ kcal.mol}^{-1}$.

However, during glucose degradation energy is transduced into ATP with a yield of 36%. Therefore, in order to perform reaction (9) a gross energy expenditure of approximately 19.5 kcal.mol^{-1} of ATP utilized should be computed. By taking into account the values of glutamine production reported in Table 2, the gross energy required for this production, under the various experimental conditions, can be calculated. The data summarized in Table 6 clearly show that the amount of energy required for the amidation of glutamate to glutamine is much lower than the change of free energy caused by the hyperammonemia syndrome.

As for the increase in the amino acid transport to the cerebral tissue, at least the following two points should be emphasized: (1) as mentioned above, the cerebral glutamate-ammonia complex seems to be a "closed system." Therefore, the increase in ammonia uptake does not seem to require any exogenous addition of amino acids, since the members of this system undergo changes which are balanced by those of the others; (2) the NH_4^+-induced changes in the arteriovenous differences in some amino acids (glutamate, glutamine, and alanine) across the brain are negligible. However, other amino acids may be carried to the brain and converted by cerebral tissue. In any case, it seems improbable

Table 6. Energy required for glutamine production

Pretreatment and treatment with	Glutamine production[a]	Energy required	Gibbs free energy[b]	Excess of free energy
	$nmol.g^{-1}.min^{-1}$	$mcal.g^{-1}.min^{-1}$		
—	43	0.84	9.10	8.26
S-adenosyl-L-methionine	54	1.05	13.57	12.52
adenosine	48	0.94	10.42	9.48

[a] Data from Table 2
[b] Data from Table 4

that the excess of cerebral free energy can be explained consistently by an increase in the amino acid transmembrane transport.

More intriguing perspectives are suggested by the problem of the changes induced by NH_4^+ in the resting transmembrane potential, which is made less negative (Table 5), 4.5-6 mV. From a clinical point of view, this ionic situation is evidenced by an increase of nervous excitability. In cerebral tissue, NH_4^+ may exchange with K^+, a K^+ efflux from the brain being induced. Therefore it may be suggested that the increased $[NH_4^+]$ and the resulting rise in $[K^+]$ may account, at least in part, for the changed metabolic rate, e.g., by activation of Na^+-K^+-stimulated adenosine triphosphatase activity.

To sum up, therefore, the experiments performed by us at cerebral level by maintaining controlled hyperammonemia conditions (0.2 mM) have shown that: (1) NH_4^+ represents the mobile molecule modulating the balanced interconversion of the members of the glutamate-ammonia system, which behaves as an integrated system; (2) the hyperammonemia syndrome induces an increase in the change of Gibbs free energy, of which an aliquot not exceeding 10% is used for the glutamate → glutamine reaction. Acute hyperammonemia induces a variation of the resting transmembrane potential, which becomes less negative. On the other hand, this change in the resting transmembrane potential can be clinically ascribed to the higher nervous excitability, which can lead to convulsions. Obviously, it should be emphasized that these conclusions are drawn from acute trials, specifically referred to the energetic and metabolic behavior of the motor cortical area. One might therefore expect that chronic trials involving other cerebral regions as well might give different results.

References

1. Alleweis, C., Magnes, J.: The uptake and oxidation of glucose by the perfused cat brain. J. Neurochem. 2, 326-336 (1968)
2. Benjamin, A.M., Quastel, J.H.: Locations of amino acids in brain slices from the rat. Tetrodotoxin-sensitive release of amino acids. Biochem. J. 128, 631-646 (1972)
3. Benjamin, A.M., Quastel, J.H.: Metabolism of amino acids and ammonia in rat brain cortex slices in vitro: a possible role of ammonia in brain function. J. Neurochem. 25, 197-206 (1975)

4. Benzi, G., Berte', F., Arrigoni, E., Manzo, L.: Study of the cerebral metabolizing activity in the newborn dog utilizing the isolated perfused brain in situ technique. J. pharm. Sci. **58**, 885-887 (1969)
5. Benzi, G., Berte', F., Crema, A., Frigo, G.M.: Cerebral drug metabolism investigated by isolated perfused brain in situ. J. pharm. Sci. **56**, 1349-1351 (1967)
6. Benzi, G., De Bernardi, M., Manzo, L., Ferrara, A., Panceri, P., Arrigoni, E., Berte'. F.: Effect of lysergide and nimergoline on glucose metabolism investigated on the dog brain isolated in situ. J. pharm. Sci. **61**, 348-352 (1972)
7. Benzi, G., Manzo, L., De Bernardi, M., Ferrara, A., Sanguinetti, L., Arrigoni, E., Berte'. F.: Action of lysergide, ephedrine, and nimergoline on brain metabolizing activity. J. pharm. Sci. **60**, 1320-1324 (1971)
8. Bergmeyer, H.U., Bernt, E.: 2-Oxoglutarate. UV spectrophotometric determination. In: Methods of Enzymatic Analysis. Bergmeyer, H.U. (ed.) New York-London: Academic Press. Second English edition, 1974, Vol. III, 1577-1580
9. Bergmeyer, H.U., Bernt, E., Möllering, H., Pfleiderer, G.: L-aspartate and L-asparagine. In: Methods of Enzymatic Analysis. Bergmeyer, H.U. (ed.) New York-London: Academic Press. Second English edition, 1974, Vol. IV, pp. 1696-1700
10. Bergmeyer, H.U., Bernt, E., Schmidt, F., Stork, H.: D-Glucose. Determination with hexokinase and glucose-6-phosphate dehydrogenase. In: Methods of Enzymatic Analysis. Bergmeyer, H.U. (ed.) New York-London: Academic Press. Second English Edition, 1974, Vol. III, pp. 1196-1201
11. Berl, S., Takagaki, G., Clarke, D.D., Waelsch, H.: Carbon dioxide fixation in the brain. J. biol. Chem. **237**, 2570-2573 (1962)
12. Beutler, H.O., Michal, G.: L-Glutamate. Determination with glutamate dehydrogenase diaphorase, and tetrazolium salts. In: Methods of Enzymatic Analysis Bergmeyer, H.U. (ed.) New York-London: Academic Press. Second English edition, 1974, Vol. IV, pp. 1708-1713
13. Cantoni, G.L.: S-adenosyl-L-methionine; a new intermediate formed enzymatically from L-methionine and adenosinetriphosphate. J. biol. Chem. **204**, 403-410 (1953)
14. Czok, R., Lamprecht, W.: Pyruvate, phosphoenolpyruvate and D-glycerate-2-phosphate. In: Methods of Enzymatic Analysis. Bergmeyer, H.U. (ed.) New York-London: Academic Press. Second English edition, 1974, Vol. III, pp. 1446-1451
15. Faulconer, A., Jr.: Correlation of concentrations of ether in arterial blood with electro-encephalographic patterns occuring during ether-oxygen and during nitrous oxide, oxygen and ether anesthesia of human surgical patients. Anesthesiology **13**, 361-366 (1952)
16. Geiger, A.: Correlation of brain metabolism and function by the use of a brain perfusion method in situ. Physiol. Rev. **38**, 1-20 (1958)
17. Geiger, A., Magnes, J.: The isolation of the cerebral circulation and the perfusion of the brain in the living cat. Amer. J. Physiol. **149**, 517-537 (1947)
18. Geiger, A., Magnes, J., Taylor, R.M., Veralli, M.: Effect of blood constituents on uptake of glucose and on metabolic rate of the brain in perfusion experiments. Amer. J. Physiol. **177**, 138-149 (1954)
19. Gilboe, D.D., Betz, A.L.: Kinetics of glucose transport in the isolated dog brain. Amer. J. Physiol. **219**, 774-778 (1970)
20. Gilboe, D.D., Betz, A.L.: Oxygen uptake in the isolated canine brain. Amer. J. Physiol. **224**, 588-595 (1973)
21. Gilboe, D.D., Andrews, R.L., Dardenne, G.: Factors affecting glucose uptake by the isolated dog brain. Amer. J. Physiol. **219**, 767-773 (1970)
22. Gilboe, D.D., Betz, L.A., Langebartel, D.A.: A guide for the isolation of the canine brain. J. appl. Physiol. **34**, 534-537 (1973)
23. Gilboe, D.D., Cotanch, W.W., Glover, M.B.: Isolation and mechanical maintenace of the dog brain. Nature (Lond.) **206**, 94-96 (1965)
24. Gilboe, D.D., Cotanch, W.W., Glover, M.B., Levin, V.A.: Changes in electrolytes, pH, and pressure of blood perfusing isolated dog brain. Amer. J. Physiol. **212**, 589-594 (1967)

25. Gilboe, D.D., Morris, B.G., Cotanch, W.W.: Blood filtration and its effect on glucose metabolism by the isolated dog brain. Amer. J. Physiol. **213**, 11-15 (1967)
26. Grassl, M.: L-Alanine. In: Methods of Enzymatic Analysis. Bergmeyer, H.U. (ed.) New York-London: Academic Press. Second English edition, 1974, Vol. IV, pp. 1682-1685
27. Greenberg, D.M.: Biological Methylation. Advanc. Enzymol. **25**, 395-431 (1963)
28. Gutmann, I., Wahlefeld, A.W.: L-Lactate. Determination with lactate dehydrogenase and NAD. In: Methods of Enzymatic Analysis. Bergmeyer, H.U. (ed.) New York-London: Academic Press. Second English edition, 1974, Vol. III, pp. 1464-1468
29. Hawkins, R.A., Miller, A.L., Nielsen, R.C., Veech, R.L.: The acute action of ammonia on rat brain metabolism "in vivo". Biochem. J. **134**, 1001-1008 (1973)
30. Hindfelt, B., Siesjo, B.K.: Cerebral effects of acute ammonia intoxication. The effect upon energy metabolism. Scand. J. clin. Lab. Invest. **28**, 365-374 (1971)
31. Krieglstein, J., Stock, R.: Comparative study of the effects of chloral hydrate and trichloroethanol on cerebral metabolism. Naunyn-Schmiedebergs Arch. Pharmacol. **277**, 323-332.(1973)
32. Kun, E., Kearney, E.B.: Ammonia. In: Methods of Enzymatic Analysis. Bergmeyer, H.U. (ed.) New York-London: Academic Press. Second English edition, 1974, Vol. IV, pp. 1802-1806
33. Lombardini, J.B., Talalay, P.: Formation, functions and regulatory importance of S-adenosyl-L-methionine. Advanc. Enzyme Regul. **9**, 349-384 (1971)
34. Lowry, O.H., Passonneau, J.V.: A Flexible System of Enzymatic Analysis. New York: Academic Press 1972, pp. 189-193
35. Lund, P.: L-Glutamine. Determination with glutaminase and glutamate dehydrogenase. In: Methods of Enzymatic Analysis. Bergmeyer, H.U. (ed.) New York-London: Academic Press. Second English edition, 1974, Vol. IV, pp. 1719-1722
36. Möellering, H.: L-Malate. Determination with malate dehydrogenase and glutamate-oxaloacetate transaminase. In: Methods of Enzymatic Analysis. Bergmeyer, H.U. (ed.) New York-London: Academic Press. Second English edition, 1974, Vol. III, pp. 1589-1593
37. Möellering, H., Gruber, W.: Determination of citrate with citrate lyase. Analyt. Biochem. **17**, 369-376 (1966)
38. Neidle, A., Berg, C.J., van den, Grynbaum, A.: The heterogeneity of rat brain mitochondria on continuous sucrose gradients. J. Neurochem. **16**, 225-234 (1969)
39. Nilsson, L., Siesjö, B.K.: The effect of anesthetics upon labile phosphates and upon extra- and intracellular lactate, pyruvate and bicarbonate concentrations in the rat brain. Acta physiol. scand. **80**, 235-248 (1970)
40. Reijnierse, G.L.A., Veldstra, H., Berg, C.J., van den: Subcellular localization of γ-aminobutyrate transaminase and glutamate dehydrogenase in adult rat brain. Biochem. J. **152**, 469-475 (1975)
41. Salganicoff, L., De Robertis, E.: Subcellular distribution of the enzymes of the glutamic acid, glutamine and γ-aminobutyric acid cycles in rat brain. J. Nuerochem. **12**, 287-309 (1965)
42. Slyke, D.D., van, Neil, J.M.: The determination of gases in blood and other solutions by vacuum extraction and manometric measurements. I. J. biol. Chem. **61**, 523-573 (1924)
43. Veech, R.L., Harris, R.L., Veloso, D., Veech, E.H.: Freeze-blowing: a new technique for the study of brain in vivo. J. Neurochem. **20**, 183-188 (1973)
44. Waelsch, H., Berl, S., Rossi, C.A., Clarke, D.D., Purpura, D.P.: Quantitative aspects of CO_2 fixation in mammalian brain "in vivo". J. Neurochem. **11**, 717-728 (1964)
45. White R.J., Albin, M.S., Verdura, J.: Isolation of the monkey brain: in vitro preparation and maintenance. Science **141**, 1060-1061 (1963)
46. Williamson, D.H., Lund, P., Krebs, H.A.: The redox state of free nicotinamideadenine dinueleotide in the cytoplasm and mitochondria of rat liver. Biochem. J. **103**, 514-527 (1967)

A Study of the Distribution of Radioactive S-Adenosylmethionine in the Experimental Animal

G.F. PLACIDI[1], G. STRAMENTINOLI[2], C. PEZZOLI[2], and G.B. CASSANO[1]

S-adenosyl-L-methionine (SAMe) is a naturally occurring molecule discovered by Cantoni (8, 9) and found in mammals. The molecular weight is 399.4 and the structural formula is as shown here:

In the living body, SAMe is synthesized by the sequence:

$$\text{methionine} + \text{ATP} \xrightarrow[\text{Mg}^{++}]{\text{E}} \text{S-adenosylmethionine} + \text{pyrophosphate} + \text{P}$$

The reaction is catalyzed by the "methionine-activating enzyme." SAMe is involved in several metabolic processes such as O-methylation, N-methylation, and S-methylation. SAMe serves as a methyl donor in the synthesis of creatine, choline, and a number of alkaloids (17), and takes part in both the synthesis and the degradation of biogenic amines (1, 4, 11, 13).

The distribution of endogenous SAMe in various body districts, and particularly in the CNS, was investigated by Baldessarini (3).

The new molecule (now available in a stabilized form from BioResearch of Milan) has already given some interesting results in preliminary clinical trials in patients with depressive syndromes (10).

In view of all these premises it was decided to investigate the distribution and tissue concentrations of exogenous SAMe administered to experimental animals, with special regard to the CNS.

Two sets of experiments were conducted. One was an autoradiographic study of whole mice and of cat brains following administration of labeled SAMe (^{14}C-methyl); the other was a quantitative assay of total radioactivity recovered in various organs of the rat.

---

[1] Department of Clinical Psychiatry, University of Pisa, Pisa (Italy).
[2] BioResearch Laboratories, Liscate (Milan, Italy).

Material and Methods

Radioactive Compound

The radioactive trace, ^{14}C-methyl-S-adenosyl-L-methionine was obtained from the Radio-chemical Centre of Amersham, England. The specific activity of the compound was 57 mCi/mmol. The working solution consisted of 50 μCi of radioactive compound dissolved in 1 ml of sulfuric acid solution, pH between 2.5 and 3.5.

Autoradiography

In a first experiment labeled SAMe was injected intravenously into eight male albino mice weighing approximately 25 g each. The dose was 1.4 μg/g, representing 0.2 μCi/g.

The animals were killed under ether anesthesia, severally after 1, 5, and 30 min and 1, 4, 12, 24, and 48 of dosing, by immersion in a mixture of hexane and solid carbon dioxide at a temperature of -70°C. The animals, evenly frozen, were embedded in carboxymethylcellulose and mounted on a special holder. Serial sagital sections (40 μ thick) of the whole animal were cut at a temperature of -10°C with a Leitz sledge microtome and then adhered on Scotch Cellulose Permanent Mending Tape N. 810 from the Minnesota Mining Manufacturing Co., USA. The slices thus prepared were kept at -10°C until they were completely dry; and brought to room temperature. The dried sections were pressed against X-ray film (Structurix, Gevaert and exposed for about 30 days, according to Ullberg's technique (19).

In another preliminary experiment the same tracer was injected intravenously into two cats, weighing approximately 1 kg each, at a dosage of 0.7 μg/g, representing 0.1 μCi/g. The two animals were killed after 12 and 24 h of the experiment, respectively, by carotid artery exsanguination. The brains were immediately removed, frozen, and sliced by the technique just described (18).

Experiment in the Rat

Fourteen male Wistar rats weighing 215 g each, fasted for 12 h were used ^{14}C-labelled SAME was injected into the tail vein at a dosage of 10 mg/kg, corresponding to 24 μCi of radioactivity. Two rats at a time were then killed by decapitation, after 15, 30, and 60 min and 4, 8, 24, and 48 h of dosing. Weighed portions of a few milligrams of the various organs were collected carefully from each animal. These, plus a 0.5-ml portion of plasma from each animal, were placed in polycarbonate capsules, which were then put in an Oxymat sampling (Intertechnique) apparatus. With this technique the sample is burned to $^{14}CO_2$, and the total radioactivity is then measured. SAMe concentrations in the tissues, plasma, and brain areas are expressed as if all the radioactivity came solely from unchanged SAMe.

Results

Autoradiography

Mice

Examination of the autoradiograms shows that from 1 to 30 min after intravenous administration of labeled SAMe most of the radioactivity is localized in the blood stream and in highly vascularized organs. High concentrations occur in the blood, lungs, liver, spleen, and kidneys, and to a lesser degree in the adrenals, skeletal muscle, intestine, and salivary glands. CNS radioactivity at this stage is confined to the vascular spaces and choroid plexuses (Fig. 1a and b).

One and 4 h after the administration the autoradiograms show high levels of radioactivity in the kidneys, spleen, liver, lungs, blood, intestine, and salivary glands; Traces of radioactivity are detectable at these observation times in the central nervous system (Fig. 1c).

Between 4 and 48 h, radioactivity decreases in the liver and gut, while it is still high in the spleen and kidneys. A progressive accumulation of tracer is detected in the myocardium and brain at 24 and 48 h (Fig. 2b and c). At that time, nearly all tissues show a radioactivity level slightly exceeding that of the blood (Fig. 2c). The following are some further details:

Central Nervous System. Radioactive SAMe crosses the blood-brain barrier of the mouse with considerable difficulty. The highest concentrations of the tracer and/or its metabolites occur in the vascular spaces and choroid plexuses. Very low levels of radioactivity were detected at 12 and 24 h of dosing, particularly around the ventricles (Fig. 2c).

Digestive Tract. High levels of radioactivity were measured in the liver. At 1 and 5 min of dosing, liver radioactivity was rather unevenly distributed and more pronounced in the more richly vascular places (Fig. 1a). At 30 min the tracer is more uniformly distributed in the liver parenchyma, and the amount of radioactivity increases to a peak at 4 h of dosing (Fig. 1b and c). Radioactivity will then gradually decrease between 12 and 48 h after administration of the tracer (Fig. 2).

In the gastric and intestinal mucosa the level of radioactivity is already high at 1 and 5 min, with further increase up to the fourth hour (Fig. 1). The salivary glands show a marked accumulation of radioactivity starting at 5 min and continuing up to the 12th hour of the experiment (Figs. 1 and 2)

Urinary Tract. The kidneys show high levels of radioactivity throughout the experiment. From 1 min to 1 h the concentration is uniform in the whole organ; but starting at 4 h there is a marked difference between the renal medulla and cortex — the latter showing far higher levels of radioactivity (Figs. 1 and 2).

Cardiovascular Apparatus. The myocardium shows levels of radioactivity definitely lower than the corresponding blood values from 1 min to 12 h after dosing (Fig. 1). At 24 and 48 h, however, radioactivity is much higher in the myocardium than in the blood going

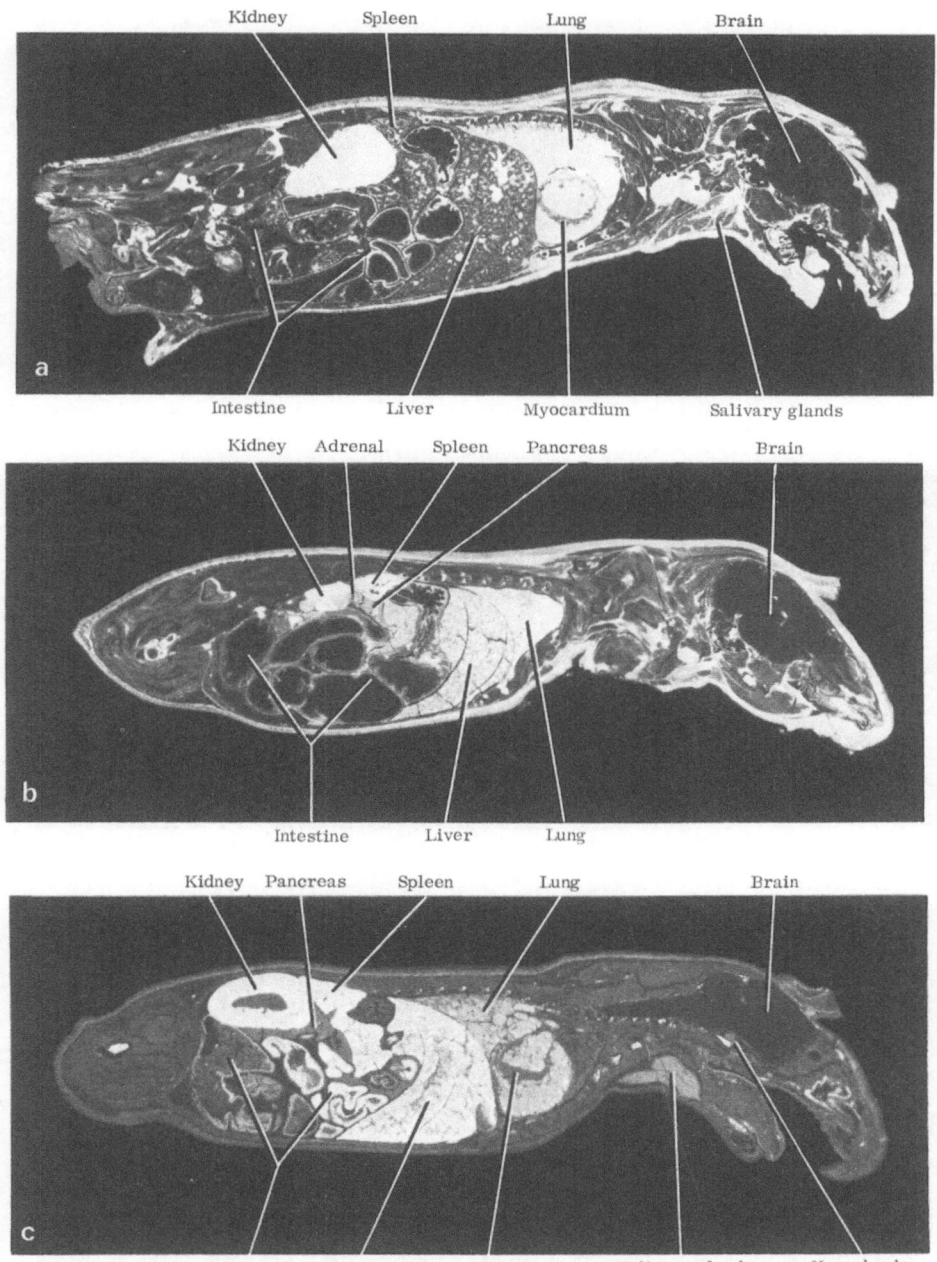

Fig. 1a-c. Whole-body mouse autoradiograms showing the distribution of radioactivity (*light areas*) at 1 min (a), 30 min (b) and 4 h (c) of intravenous dosing with radioactive SAMe

104

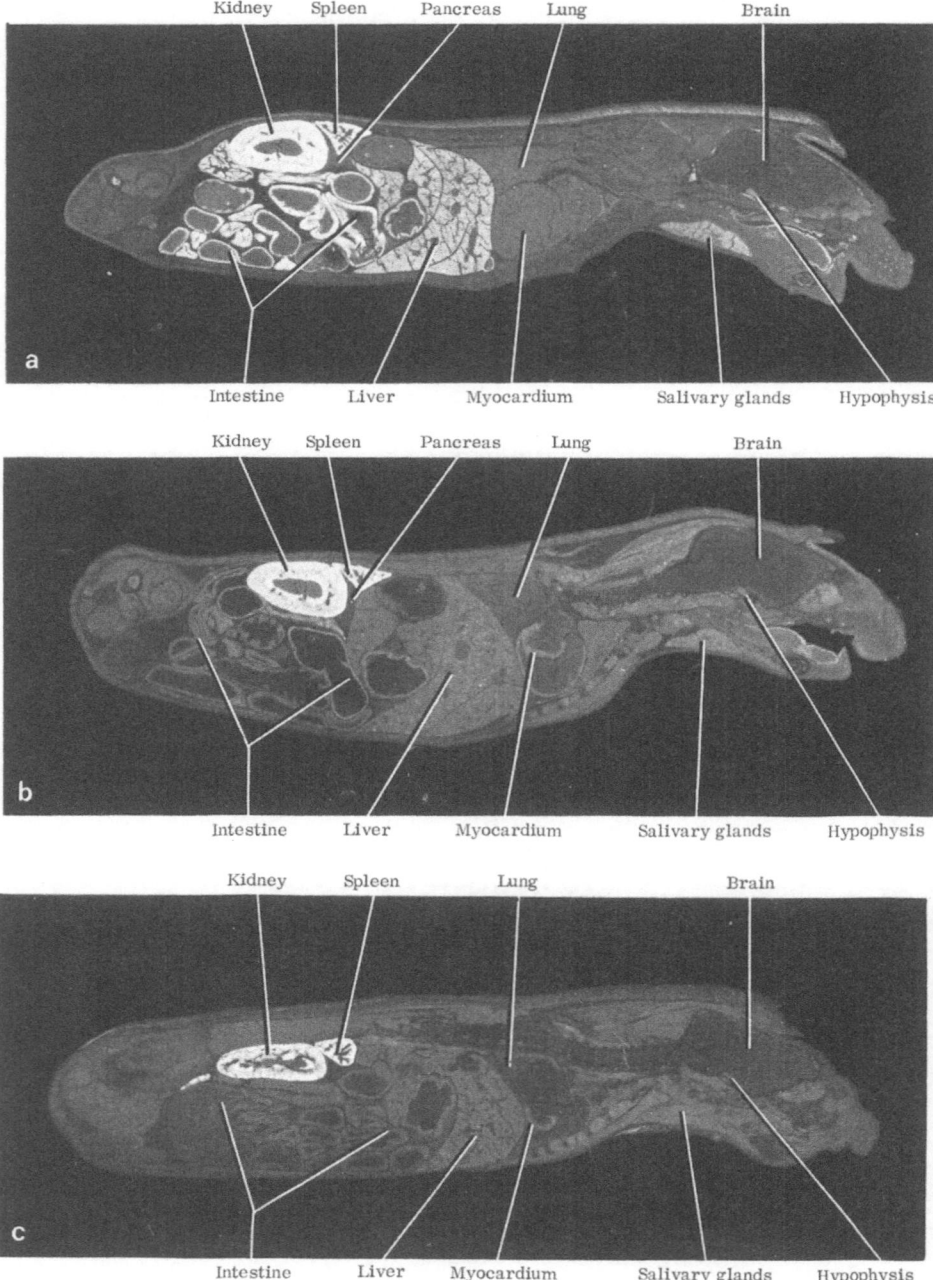

Fig. 2a-c. Whole-body mouse autoradiograms showing the distribution of radioactivity (*light areas*) at 12 h (a), 24 h (b) and 48 h (c) of intravenous dosing with radioactive SAMe

through the heart chambers. In the blood, radioactivity continues high for the first
4 h, then decreases gradually (Fig. 2).

Fat, Skeletal Muscle, and Skin. Adipose tissue shows negligible amounts of radioactivity
at all times. Skeletal muscle shows less radioactivity than the circulating blood up to
12 h after dosing. Between 24 and 48 h, while still very low in absolute, muscle radioactivity exceeds blood readings (Fig. 2). In the skin, the highest measurements of radioactivity are elicited early in the experiment.

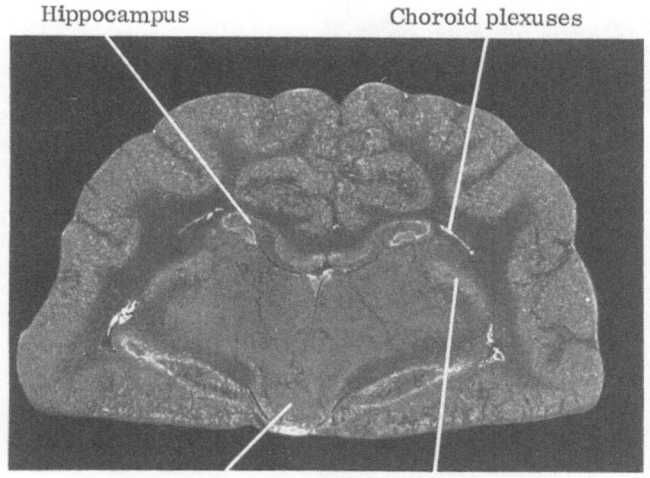

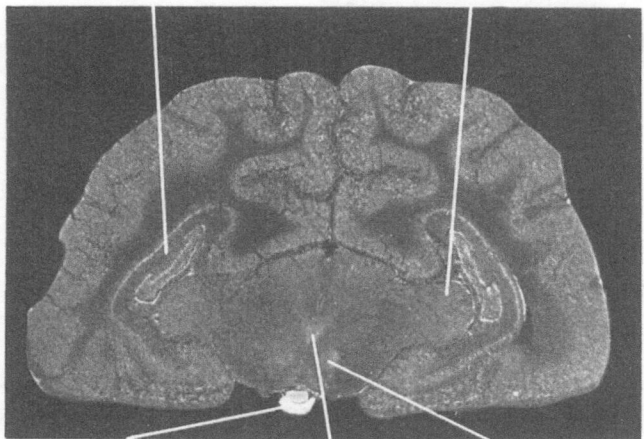

Fig. 3. Cat brain auto-
radiograms showing
the distribution of
radioactivity (*light
areas*) in two coronal
sections at 12 h of in-
travenous dosing with
labeled SAME

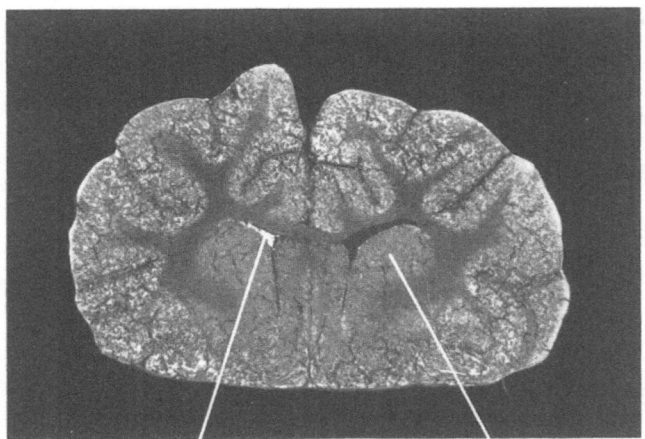

Fig. 4. Cat brain auto-
radiograms showing
the distribution of
radioactivity (*light
areas*) in coronal sec-
tions at 24 h of intra-
venous dosing with
labeled SAMe

Choroid plexuses   Caudate nucleus

Lateral geniculate body   Medial geniculate body

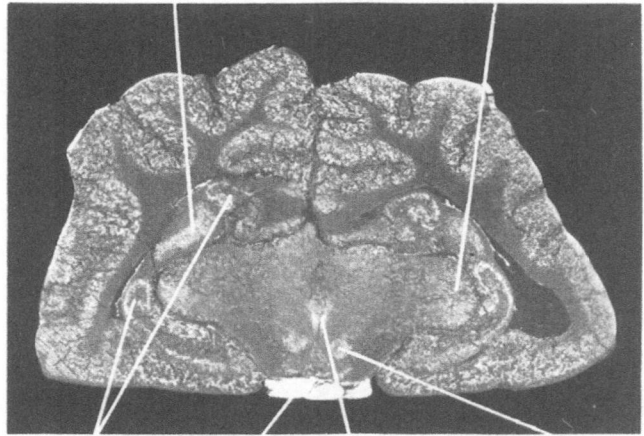

Hippocam.   Hypophysis   Oculomotor n.   Red nucleus

Other Organs. The spleen shows very high concentrations throughout the experiment, from 5 min to 48 h after dosing, and predominantly in the red pulp (Figs. 1 and 2). Likewise, bone tissue appropriates a lot of radioactivity and retains it from 5 min to 48 h after dosing (Figs. 1 and 2). The pancreas shows a peak of radioactivity at 30 min and 1 h, followed by rapid dissipation (Fig. 1). The adrenals show increasing uptake up to the fourth hour, then steady levels exceeding the corresponding blood readings up to 48 h.

Finally, the hypothysis shows a high level of radioactivity as early as 1 min after dosing; the level increases further up to the fourth hour and then tapers off at later readings. The testes show only a faint radioactivity at 1 and 5 min of dosing, then practically none (Fig. 1).

Legend see p. 106

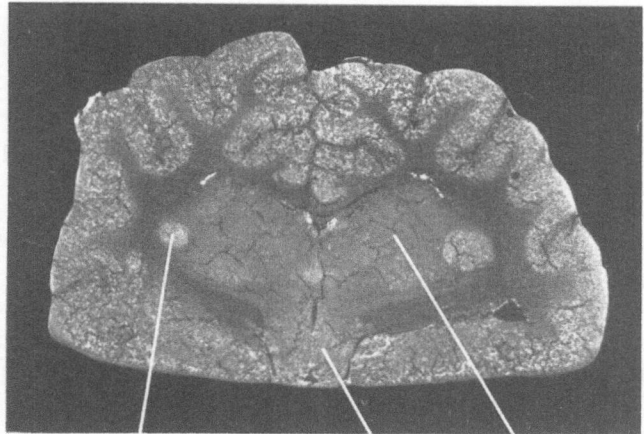

Lateral geniculate body Hypothalamus Thalamus

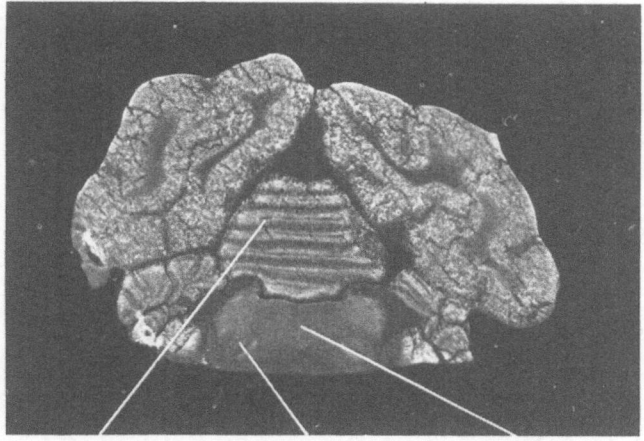

Cerebellar cortex Upper olivary n. Reticular form.

Cats

In this set of experiments we examined the distribution of radioactive SAMe at 12 and 24 h after administration of the tracer. This was done after we had found that the highest readings of radioactivity were elicited in the mouse brain at 24 and 48 h of dosing. In our cats, both at 12 and at 24 h, most of the brain radioactivity was detected in the choroid plexuses, brain cortex, caudate nucleus, pyramidal layer of the hippocampus, some thalamic nuclei such as the lateral and medial geniculate bodies, and mesencephalic nuclei such as the oculomotor and red nucleus. The hypothalamus showed about the same amount of radioactivity as the brain cortex (Figs. 3 and 4).

The general pattern of distribution of brain radioactivity was the same at 12 and at 24 h of dosing, although absolute values were greater at 24 h.

Organ Radioassay

Rats

The data in Table 1 show that the distribution of radioactivity in the rat duplicates the pattern seen in the autoradiography studies of mice. Plasma radioactivity dropped sharply after an early peak, this fact reflecting the short half-life of SAMe administered intravenously. At the same time, radioactivity accumulated in the liver, spleen, kidneys, and adrenals. In the liver, in fact, radioactivity readings continued high even at 24 h, this being the only notable difference from what was observed in the mice, possibly due to the different dosages used in the two species, or perhaps to different ways of metabolizing the product. Incidentally, the LD_{50} value of SAMe is about four times greater in rats than mice, again suggesting different metabolisms in the two species.

As for brain radioactivity (Table 2), the rats showed low values compared to other organs, as did the mice. Peak readings were elicited from the rat brain at 24 h of dosing.

Table 1. Tissue and plasma levels of labeled SAMe in the adult rat. Values expressed as $\mu g/g$ or $\mu g/ml$

Tissue	15 min	30 min	60 min	4 h	8 h	24 h	48 h
Liver	4.02	4.31	7.47	13.1	13.44	13.57	10.5
Spleen	3.15	2.80	2.96	8.1	6.25	6.72	5.45
Adrenals	4.26	6.47	8.46	11.7	10.86	10.93	8.4
Kidneys	174.5	297.5	310.0	176.7	111.0	62.0	43.7
Plasma	18.6	9.43	5.2	5.3	5.3	3.15	2.33

(Total radioactivity elicited is regarded as coming exclusively from labeled SAMe).

Table 2. Concentrations of labeled SAMe in various brain areas of the adult cat. Values expressed as $\mu g/g$

Brain areas	15 min	30 min	60 min	4 h	8 h	24 h	48 h
Cerebellum	0.57	0.64	0.88	3.1	3.45	3.67	2.20
Spinal cord	0.68	0.67	0.72	2.51	2.55	3.99	2.34
Med.oblongata	0.88	0.78	0.85	2.32	2.88	2.67	1.94
Hypothalamus	0.74	1.07	1.53	2.45	3.02	3.20	2.11
Mesencephalon	0.36	0.42	0.58	1.96	2.21	2.86	1.82
Cortex	0.59	0.56	1.10	1.82	2.10	2.36	1.72
Hypophysis	5.60	5.15	5.82	19.5	11.35	10.32	6.17

(Total radioactivity elicited is regarded as coming exclusively from labeled SAMe).

Discussion

Autoradiography studies in the mouse showed that following a single intravenous injection of labeled SAMe the immediate distribution of the substance in the animal's body closely reflected the vascularity and blood flow of the various organs (Fig. 1a): at 1 min of dosing, most of the radioactivity was detected in the blood stream and in the lungs, adrenals, vascular spaces of the liver, intestinal walls, and spleen. All other organs and tissues showed low radioactivity — less, at any rate, than that of the circulating blood. On the other hand, the steady accumulation of radioactivity in the liver, kidneys, intestine, and salivary glands after the decrease of plasma radioactivity levels (Fig. 2a) suggests the role of these organs in the metabolic processing and excretion of the substance; this is particularly true of the kidneys, where radioactivity was still very high at 48 h. A peculiar aspect of SAMe was its marked uptake by the spleen, with steadily high radioactivity readings from 5 min to 48 h.

Also interesting is the late accumulation of radioactivity in the myocardium (between 24 and 48 h after dosing), suggesting the existence of an uptake mechanism in this tissue rich in adrenergic neuron terminals (14).

Labeled SAMe crossed the blood-brain barrier only slowly and sparingly. After some relatively high readings in the first few minutes, reflecting the presence of radioactive blood in the large vascular spaces and choroid plexuses, brain radioactivity rose slightly and gradually to reach a maximum after 24 h of dosing. The mouse brain being too small to reveal regional differences of tracer concentration, we studied this aspect of the problem in a separate experiment of cat brain autoradiography.

The results of these experiments indicate a distribution of labeled SAMe closely comparable to that of some tricyclic psychotropic drugs such as chlorpromazine, amitripyline and imipramine (5-7), with preferential accumulation in the cortex, caudate nucleus, geniculate bodies, basal nuclei, brain stem nuclei, and cerebellar cortex. While similar in pattern, however, this distribution was achieved with SAMe in a much longer time than with the tricyclic compounds just named, in keeping with the slow penetration of the tracer into the CNS. Also, it seems that SAMe probably reaches the brain via the cerebrospinal fluid (CSF), as indicated by the high concentration of radioactivity in the choroid plexuses at all observation times and by the accumulation of radioactivity in the hippocampus — which, according to several authors, reflects the slow diffusion of psychotropic drugs (5) and other substances (2, 12, 16) from the CSF.

The hypophysis showed high concentrations of radioactivity at all observation times, buth this finding lacked specificity, since the most diverse substances seem to concentrate in that organ because of its rich vascularity and because the hypophysis is outside the blood-brain barrier.

Conclusions

Our experimental findings indicate that labeled SAMe crosses the blood-brain barrier very slowly but progressively, peak concentrations beyond the barrier being invariably lower than in other organs included in this study. This, however, should not be construed as meaning that SAMe has little or no pharmacologic action in the CNS: other drugs that are very active in the CNS occur at very low concentrations in the brain — suffice it to mention reserpine (15).

Acknowledgments. This study was partly supported by a CNR grant.

We owe Mr. Mauro Brogiotti special thanks for his competent technical assistance.

References

1. Axelrod, J.: O-methylation of epinephrine and other catechols in vitro and in vivo. Science **126**, 400 (1957)
2. Bakay, L.: Studies on blood-brain barrier with radioactive phosphorus. Arch. Neurol. Psychiat. (Chicago) **71**, 673 (1954)
3. Baldessarini, R.J., Kopin, I.J.: S-adenosylmethionine in brain and other tissues. J. Neurochem. **13**, 769 (1966)
4. Brown, D.D., Axelrod, J., Tomchick, R.: Enzymatic N-methylation of histamine. Nature (Lond.) **183**, 680 (1959)
5. Cassano, G.B., Hansson, E.: Autoradiographic distribution studies in mice with ^{14}C-imipramine. Int. J. Neuropsychiat. **2**, 269 (1966)
6. Cassano, G.B., Sjöstrand, S.E., Hansson, E.: Distribution of ^{35}S-chloropromazine in cat brain. Arch. int. Pharmacodyn. **156**, 48 (1965)
7. Cassano, G.B., Sjöstrand, S.E., Hansson, E.: Distribution of ^{14}C-labelled amitryptyline in the cat brain. Psychopharmacologia (Berl.) **8**, 12 (1965)
8. Cantoni, G.L.: The nature of the active methyl donor formed enzymatically from L-methionine and adenosintriphosphate. J. Amer. Chem. Soc. **74**, 2942 (1952)
9. Cantoni, G.L.: S-Adenosylmethionine, a new intermediate formed enzymatically from L-methionine and adenosintriphosphate. J. biol. Chem. **204**, 403 (1953)
10. Fazio, C., Andreoli, V., Agnoli, A., Casacchia, M., Cerbo, R.: Effetti terapeutici e meccanismo d'azione della S-adenosil-L-metionina nelle sindromi depressive. Minerva Med. (Torino) **64**, 1515 (1973)
11. Kirschner, N., Goodall, M.: The formation of adrenaline from noradrenaline. Biochim. biophys. Acta **24**, 658 (1957)
12. Lee, J.C., Olszewski, J.: Penetration of radioactive bovine albumin from cerebrospinal fluid into brain tissue. Neurology (Minneap.) **10**, 814 (1960)
13. Lindhal, K.M.: The histamine methylating enzyme system in liver. Acta Physiol. scand. **49**, 114 (1960)
14. Masuoka, D.T., Placidi, G.F.: Uptake of ^{3}H-norepinephrine by fluorescent nerves of the heart. J. Histochem. Cytochem. **18**, 660 (1970)
15. Placidi, G.F., Ciccone, G., Moschini, A., Gliozzi, E., Cassano, G.B.: Autoradiographic distribution studies with ^{3}H-reserpine in mice. J. Nucl. Biol. Med. **16**, 32 (1972)
16. Roth, L.J., Schoolar, J.C., Barlow, C.F.: Sulphur-35-labelled acetazolamide in cat brain. J. Pharmacol. exp. Ther. **125**, 128 (1959)
17. Shapiro, S.K., Schlenk, F.: Transmethylation and methionine biosynthesis. Chicago Press Publ., Chicago, Illinois, p. 26, 1965
18. Ullberg, S.: Studies on the distribution and fate of ^{35}S-labelled benzyl-penicillin in the body. Acta radiol. (Stock), suppl. **118**, 1954

Concentrations of S-Adenosyl-L-Methionine in the CNS of Rats After Intramuscular and Intravenous Administration

G. STRAMENTINOLI[1], E. CATTO[1], and S. ALGERI[2]

Recent clinical studies have demonstrated the therapeutic effectiveness of S-adenosyl-L-methionine (SAMe) in the treatment of depressive syndromes (1).

Since SAMe is intended for parenteral administration in man, we considered it of interest to find out how much of it could be recovered in the brain of experimental animals (rats) after a single intramuscular or intravenous injection of the substance.

Previous studies involving whole-body autoradiography of mice treated with labeled SAMe at a dosage of 1.4 $\mu g/g$ had shown that the radioactivity bound to the substance (or its metabolites) penetrated the CNS only slowly and in sparing amounts. As the dosages used in those experiments was extremely low, we decided to explore the passage of SAMe into the rat brain with larger doses.

Material and Methods

We used male Sprague-Dawley rats weighing 250 ± 10 g, supplied by the C. River Company and maintained on a standard diet with free access to food and water.

The labeled product, ^{14}C-methyl-SAMe, was supplied by the Radiochemical Centre of Amersham, England, from plain SAMe in the form of a stable salt supplied by the BioResearch Company of Liscate (Milan, Italy).

The animals received the labeled product dissolved in a phosphate buffer solution, at a dosage of 100 mg/kg both for the intramuscular and for the intravenous treatment groups; the volume of injection was 0.2 ml/100 g animal weight in all cases.

The rats were killed by decapitation at 15, 30, 60, and 120 min of intramuscular dosing, and at 5, 15, 30, 60, and 120 min of intravenous dosing. The brains were immediately collected and washed in cold physiologic saline. In the intravenous treatment group the brain and cerebellum were separated and assayed separately for SAMe content. The brains were immediately frozen in solid carbon dioxide, weighed, and homogenated in 10 volumes of 10% trichloroacetic acid in 0.05 N hydrochloric acid solution. After centrifugation for 10 min at $+4^{o}$C, 1 ml portions of the clear supernatant were used for assaying SAMe by the method of isotope dilution described by Baldessarini and Kopin (3).

---

[1] BioResearch Company, Department of Biochemistry, Liscate (Milan), Italy.
[2] Mario Negri Institute of Pharmacological Research, Milan (Italy).

Results and Discussion

Table 1 shows the values of SAMe assays in the whole brains at different time intervals after a single intramuscular dose of 100 mg/kg. SAMe contents in the whole brain were significantly greater than the control values at 15, 30, and 60 min of dosing. The highest concentration was found at 60 min, with a 45.9% increase over control values. The values found at 2 h were back to control levels.

Table 1. SAMe concentrations in the brain of rats after single intramuscular doses of 100 mg/kg. Listed values represent the mean ± standard error for five animals in each group

Time after treatment (min)	SAMe assay µg/g
0 (control)	9.33 ± 0.26
15	10.85 ± 0.27*
30	11.25 ± 0.52*
60	13.62 ± 0.17*
120	9.45 ± 0.31

*$P < 0.01$

Statistical significance for each group versus the controls was assessed by Duncan's new multiple range test.

Earlier studies of the pharmacokinetics of SAMe in the plasma of rats after intramuscular dosing (5) had shown a peak plasma concentration of the product at 5 and 10 min of dosing, followed by rapid dissipation of the substance from the blood stream. This indicates a very short half-life of SAMe in the plasma of rats (37.8 min).

Peak SAMe levels in the brain are achieved at 60 min after intramuscular administration, when plasma levels are already back to or near control values (5).

Thus it appears certain that the SAMe assayed in the brain at 1 h does not represent SAMe circulating in the brain vasculature at that time.

In another set of experiments we assayed SAMe separately in the whole brain and whole cerebellum of rats treated with a single intravenous dose of 100 mg/kg. The results (Table 2) indicate that peak SAMe concentrations are obtained in the brain 5 min after injection, the increase being more than 100% over basal values. SAMe assays continue significantly higher than control values at 30 min and are still about 24% above the norm at 2 h. Comparable results were obtained from the cerebellum, which as we said was assayed separately. From 5 to 60 min after dosing the increment of SAMe content in the cerebellum was statistically significant in all animals, with mean increases of 100% over control values at 5 min and 29.8% at 120 min.

Data published by Baldessarini (2) show that the administration of methionine (a SAMe precursor) by intraperitoneal injection to rats at a dosage of 666 µmol/kg (representing 2.6 times the SAMe dosage adopted in our experiments) produced a 30.4% increase of the brain SAMe assay at 30 min of dosing.

Table 2. SAMe concentrations in the brain and cerebellum of rats after single intravenous doses of 100 mg/kg. Listed values represent the mean ± standard error for five animals in each group

Time after treatment (min)	SAMe assay μg/g	
	Brain	Cerebellum
0 (control)	9.35 ± 0.64	11.40 ± 0.42
5	19.76 ± 1.37*	23.51 ± 2.29*
15	13.77 ± 1.28*	17.20 ± 1.18*
30	12.99 ± 0.38*	16.98 ± 0.31*
60	11.98 ± 0.93	16.80 ± 0.84*
120	11.60 ± 0.71	14.80 ± 0.56

*$P < 0.01$

Statistical significance for each group versus the controls was assessed by Duncan's new multiple range test.

Our own previous studies of autoradiography in the mouse (4) showed that labeled SAMe administered in relatively small doses crossed the blood-brain barrier only slowly and sparingly. In this regard we must also bear in mind that autoradiography does not afford chemical identification of a test product, so that we do not know whether the measured radioactivity was from SAMe as such or from some metabolite. The scant brain penetration of SAMe observed in those experiments may be taken to mean that the extremely small amounts of product administered were rapidly metabolized peripherally before they had a chance of reaching the brain. Later experiments in which we treated rats with intravenous SAMe at a dosage of 10 mg/kg confirmed that the substance penetrated the brain rather poorly.

The data of this present investigation, however, show that following administration of relatively large doses of SAMe, either by intramuscular or intravenous injection, the substance accumulates in brain tissue to exceed control values by as much as 100%.

References

1. Agnoli, A., Andreoli, V., Casacchia, M., Cerbo, R.: Effect of S-Adenosyl-L-Methionine (SAMe) upon depressive symptoms. J Psychiat. Res. **13**, 43-54 (1976)
2. Baldessarini, R.J.: Alterations in tissue levels of S-Adenosyl-L-Methionine. Biochem. Pharmacol. **15**, 741-748 (1966)
3. Baldessarini, R.J., Kopin, I.J.: S-Adenosyl-L-Methionine in brain and other tissues. J. Neurochem. **13**, 769-777 (1966)
4. Placidi, G.F., Stramentinoli, G., Pezzoli, C., and Cassano, G.B.: A study of the distribution of radioactive S-Adenosylmethionine in the experimental animal. This volume (Chapter 8) p. 100 ff.
5. Stramentinoli, G., and Catto, E.: Pharmacokinetic studies of S-Adenosyl-L-Methionine (SAMe) in several animal species. Pharmacol. Res. Commun. **8**(2), 211-218 (1976)

The Subcellular Distribution of S-Adenosyl-L-Methionine (SAMe) in the Rat Brain

V.M. ANDREOLI[1], F. MAFFEI[1], and G.C. TONON[2]

In order to further our understanding of transmethylation processes (2) in the CNS, which are now thought to play a role in the biochemistry of behavior through an action on the mediators of such behavior, namely the biogenic amines, we have done some work seeking correlations between the subcellular distribution of S-adenosyl-L-methionine (SAMe) and that of catechol-O-methyltransferase (COMTs), the latter being the best-known transmethylases. The premises of this investigation were as follows:

(a) SAMe is one of the methyl donors occurring in the CNS, where it shows a characteristic topographic distribution. According to Baldessarini and Kopin (3) the highest concentrations of SAMe occur in the basal nuclei of the brain (14.2 μg/g of tissue) and the lowest occur in the cerebellum (10.8 μg/g of tissue), with a mean value of 12.5 μg/g for the encephalon as a whole. Such values are of the order of about one-fourth those detected in the adrenal glands (48.0 ± 10.0 μg/g), these being the organs that contain more SAMe than any other in the body. The distribution of SAMe can be deduced indirectly from that of the "methionine-activating enzyme," which governs the synthesis of SAMe. According to Volpe and Laster (11) the assay of brain tissues for methionine-activating enzyme (MAE) give the following concentrations in decreasing order: hypophysis 142.0 ± 11,2; cerebellum 131.0 ± 4.8; hypothalamus 80.2 ± 3.2; cerebral parietal cortex 79.7 ± 2.6; hippocampus 67.5 ± 3.0; thalamus 59.8 ± 2.7; white parietal matter 53.0 ± 2.8. These data, unlike Baldessarini and Kopin's (3), indicate higher concentrations of MAE in the cerebellum.

(b) The process of transmethylation presupposes the action of a specific enzyme; among SAMe-dependent enzymes we find the COMTs, which play a role in the inactivation of catecholamines (1).

With such premises, one would expect to find SAMe in the same places where one finds COMTs; and if such similarity of distribution exists, then one might expect that a change of SAMe concentrations (or rather, of the SAMe → S-adenosylhomocysteine equilibrium) would interfere with the transmethylation process and so, through involvement of the central mediators, produce alterations of behavior.

---

[1] Neuropsychiatric Hospitals, Verona (Italy).
[2] Department of Pharmacology, University of Milan (Italy).

Material and Methods

The experimental animals were male Sprague-Dawley rats weighing between 250 and 300 g, housed at constant temperature and relative humidity and maintained on a standard laboratory diet for rodents. The same was assayed on subcellular fractions from brain homogenates by the Gray and Whittaker fractionating technique (7, 9).

One group of rats served as untreated controls; the other group was treated with an interventricular injection of 40 μl of labeled SAMe (^{3}H-methyl) with 29 mCi/mg specific activity; the experimental design was Glowinski's (6) as modified by Schamberg (10). As an internal standard, exactly the same amount of labeled SAMe was added to brain homogenates from untreated animals. All the rats were then killed by decapitation; their brains were collected, stored at +4°C, and then homogenated in 0.32 M saccharose (12). With this technique we investigated, in addition to the endogenous distribution, the distribution of exogenous SAMe administered by intraventricular injection.

Endogenous SAMe was assayed by the method of Baldessarini (3) as modified by Matthysse (8). Radioactive samples were read after freezing and thawing of subcellular fractions, followed by deproteinization. Readings were taken with a liquid scintillation counter (Packard Tricarb Model 3385) at a neutral pH in Instagel.

Recovery with the various assay methods, calculated for the primary fractions on the homogenate as a whole, and for the secondary fractions on P_2, was in the region of 90% (9).

Results and Discussion

Table 1 shows the distribution of endogenous SAMe in the primary and secondary fractions. As can be seen, the SAMe assay in the crude homogenate was 9.29 ± 1.28 μg/g of tissue, only slightly below published data (3). Most of the SAMe was fairly evenly divided between fraction P_2 representing the synaptosomes (32.66%) and the supernatant fraction S_3 (37.19%). In order to detect a possible specific binding to subcellular structures, we calculated the relative specific activity (RSA) (9) of each fraction, i.e., the ratio of percentage activity of a given fraction to percentage protein content of the same fraction. The data in Table 1 show that RSA was significant (2.211) only in fraction S_3: in other words, only this fraction showed a specific ratio of SAMe to protein content.

As for the secondary fraction, most of the SAMe was in fraction C (mitochondria); this was also the only fraction showing a significant RSA (1.966). This last finding confirms the indication emerging from fraction P_2, namely the lack of specific localization in the synaptosomes, of which fraction B represents an enriched subfraction.

Table 2 shows the distribution of COMTs. In good agreement with data from other sources (4, 9), we found the enzyme located in the soluble S_3 fraction (43.11%; RSA 2.564); in the other primary and secondary fractions there were no RSA values indicating selective localization.

Thus a comparison of Tables 1 and 2 shows quite clearly that the subcellular localization of SAMe duplicates that of COMTs; in turn, this suggests that the two substances are dynamically interconnected and this is why they occur with the same topographic

Table 1. Distribution of endogenous SAMe in the primary and secondary fractions of rat brain homogenates

Fractions	SAMe	Proteins	RSA
H (μg/g tissue)	9.29 ± 1.28	118.32 ± 2.95	–
P_1 (% H)	23.47 ± 0.53	37.41 ± 3.78	0.627
P_2 (% H)	32.66 ± 1.79	39.68 ± 2.03	0.823
P_3 (% H)	7.35 ± 0.32	7.08 ± 1.04	1.038
S_3 (% H)	37.19 ± 1.67	16.82 ± 0.90	2.211
A (% P_2)	28.00 ± 3.83	27.40 ± 0.28	1.021
B (% P_2)	28.73 ± 3.92	54.32 ± 3.68	0.522
C (% P_2)	43.32 ± 2.72	22.04 ± 7.05	1.966

Table 2. Distribution of COMTs in the primary and secondary fractions of rat brain homogenates

Fractions	COMTs	Proteins	RSA
H (mμM/30'/g tissue)	11.76	118.32 ± 2.95	–
P_1 (% H)	22.77 ± 4.04	37.41 ± 3.78	0.749
P_2 (% H)	28.06 ± 2.05	39.68 ± 2.03	0.650
P_3 (% H)	9.05 ± 1.13	7.08 ± 1.04	1.049
S_3 (% H)	43.11 ± 2.55	16.82 ± 0.90	2.564
A (% P_2)	30.89 ± 1.15	27.40 ± 0.28	1.127
B (% P_2)	49.88 ± 2.03	54.32 ± 3.68	0.918
C (% P_2)	19.18 ± 0.52	22.04 ± 7.05	0.870

distribution. For both substances the synaptosome fraction (where the catechols are located) is nonspecific; this suggests that their action on the catecholamine substrate can take place only outside the synapse endings. Thus SAMe is present in the soluble fraction and is not particulated. Unlike the COMTs, SAMe appears to have a fairly significant secondary location in the mitochondrial fraction (secondary fraction C of primary fraction P_2). This finding suggests a possible correlation with mitochondrial ATP and may reflect the synthesis of SAMe in the mitochondria, as suggested by Baldessarini (3).

Let us now take a look at the preliminary results obtained in rats given radioactive SAMe intraventricularly. The purpose of these experiments was simply to find out whether or not the distribution of endogenous SAMe was specific, i.e., restricted to some subcellular structures and not to others. To this end, the brains of rats so treated were collected 30 min after injection and processed in exactly the same way as those assayed for endogenous SAMe. Figure 1 shows the percentage variations of radioactivity in the various primary and secondary fractions at 30 min of dosing. From these experiments it emerges that the assay of labeled SAMe in soluble fraction S_3 is 67.23 % at zero time –

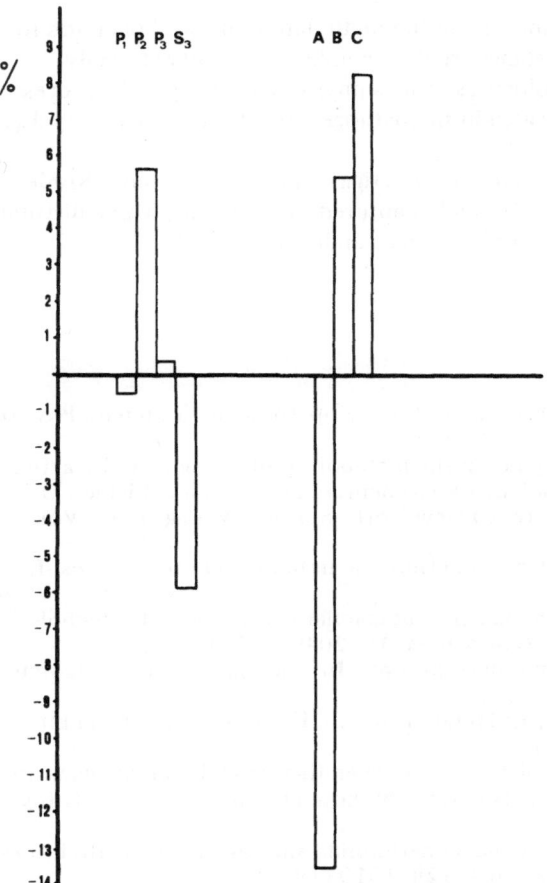

Fig. 1. Percentage variation of radioactivity (DPM/g of tissue) 30 min after intraventricular administration of labeled SAMe in the rat

much higher, therefore, than the 37.19% of endogenous SAMe; but at time 30 min the assay drops to 61.82% in fraction S_3 and at the same time it rises in fraction P_2. This indicates that the distribution of administered SAMe takes place very slowly, with only 5.41% of supernatant radioactivity being distributed at 30 min, as opposed to a theoretical 30.04%. Still, the timing used in the experiment is enough to show that a slow distribution does occur, with shifting of radioactivity from the supernatant to the P_2 fraction, and within the latter from subfraction A (rich in myelin) to subfractions B and C, representing respectively the synaptosomes and the mitochondria — where, however, the values are still far short of physiologic levels: namely 21.39 instead of 28.73 in subfraction B, and 17.38 instead of 43.32 in subfraction C.

Experiments with different timings are under way in our laboratory. The data presented here, however, indicate that the distribution follows a similar pattern to the values found for the subcellular distribution of endogenous SAMe.

From our results, therefore, it may be inferred that SAMe, showing as it does a CNS distribution analogous to that of COMTs, probably represents the methyl donor utilized by COMTs in their biochemical operations, from the methylation of catecholamines to that of exogenous phenethylamine substances such as amphetamine. More broadly, SAMe might easily be involved in pathologic as well as physiologic methylation processes — the pathologic variety being advocated in the pathogenesis of certain neuropsychiatric syndromes (5).

One result clearly emerging from this present investigation is that exogenous SAMe delivered into the brain ventricles of rats is slowly captured by CNS cells, with a distribution pattern that tends to duplicate the distribution of endogenous SAMe.

References

1. Axelrod, J.: Metabolism of epinephrine and other sympathomimetic amines. Physiol. Rev. **39**, 751 (1959)
2. Axelrod, J.: Methyltransferase enzymes in the metabolism of physiologically active compounds and drugs. In: Handbook of Experimental Pharmacology. Eichler, O., Farah, A., Herken, H., Welch, A.D. (eds.) New York: Springer Verlag, 1971, Vol. XXVIII (2), p. 610
3. Baldessarini, R.J., Kopin, I.J.: S-adenosylmethionine in brain and other tissues. J. Neurochem. **13**, 769 (1966)
4. Broch, O.J., Fonnum, O.J.: The regional and subcellular distribution of catechol-O-methyltransferase in rat brain. J. Neurochem. **19**, 2049 (1972)
5. Friedhoff, A.J.: Biogenic amines and schizophrenia. In: Biological Psychiatry. Mendels, J. (ed.) New York: Wiley 1973, pp. 113-129
6. Glowinski, J., Kopin, I.J., Axelrod, J.: Metabolism of (H^3) norepinephrine in rat brain. J. Neurochem. **12**, 25 (1965)
7. Gray, E.G., Whittaker, V.P.: The isolation of nerve endings from brain: an electron microscopic study of the cell fragments derived by homogeneization and centrifugation. J. Anat. **96**, 79 (1962)
8. Matthysse, S., Baldessarini, R.J.: S-adenosylmethionine and catechol-O-methyltransferase in schizophrenia. Amer. J. Psychiat. **128**, 1310 (1972)
9. Ross, L.L., Andreoli, V.M., Marchbanks, R.M.: A morphological and biochemical study of subcellular fractions of the guinea pig spinal cord. Brain Res. **25**, 103 (1971)
10. Schamberg, S.M., Schildkraut, J.J., Kopin, I.J.: The effects of phentobarbital on the fate intracisternally administered norepinephrine H^3. J. Pharmacol. exp. Ther. **157**, 311 (1967)
11. Volpe, J.J., Laster, L.: Trans-sulphuration in primate brain regional distribution of methionine-activating enzyme in the brain of the Rhesus monkey at various stages of development. J. Neurochem. **17**, 413 (1970)
12. Whittaker, V.P., Dowe, G.H.C.: The effects of homogeneization conditions on the subcellular distribution in brain. Biochem. Pharmacol. **14**, 184 (1965)

On Lowering S-Adenosylmethionine

S. MATTHYSSE

L-Dopa produces a marked lowering of brain S-adenosylmethionine (8), but nicotin-amide, although also a methyl group acceptor, does not (3). It would be interesting to have a theoretical explanation for this difference in activity, especially if it would help in the selection of new agents which might have S-adenosylmethionine lowering activity.

Let us adopt the Michaelis-Menten formulation of enzyme activity (V) in terms of substrate concentration (S):

$$V = \frac{V_{max}}{1 + K_m/S}$$

Let V_0, S_0 represent conditions normally prevailing in the tissue. If a large excess of sub-strate is added (as would be the case when pharmacologic doses of methyl acceptors are used to lower S-adenosylmethionine), V tends toward V_{max}. The maximum possible in-crease in methyl group utilization by adding a given acceptor is:

$$V_{max} - V_0 = V_{max}(1 - \frac{1}{1 + K_m/S_0}) = \frac{V_{max}}{1 + S_0/K_m}$$

An alternative formulation is:

$$V_{max} - V_0 = V_0 (\frac{V_{max}}{V_0} - 1) = V_0 K_m/S_0$$

We can attempt to estimate K_m, S_0, and either V_{max} or V_0 from the available data on each methyl group acceptor. In general, V_{max} will be estimated from in vitro data on an isolated and perhaps partially purified enzyme, whereas V_0 represents normal in vivo ac-tivity. For the purpose of predicting the increment in methyl group transfer that can be effected by in vivo administration of an acceptor, the second method (using V_0) is clearly preferable, but data may not always be available.

Let us now consider catechol-O-methyltransferase, with norepinephrine as substrate, and nicotinamide methyltransferase. We will use the first method of estimation with rat liver as the tissue (data for nicotinamide on brain not being available).

Using these estimates, the ratio of the S-adenosylmethionine lowering capacity of nor-epinephrine to that of nicotinamide is approximateiy 300:1. When L-dopa is used as the methyl acceptor, the corresponding endogenous substrate determining V_0 is primarily norepinephrine, although a correction should be made for the other substrates of COMT

Mailman Research Laboratories, McLean Hospital, Belmont, Mass. 02178, USA.

Table 1.

Acceptor	V_{max}	$S_0 (\mu M)$	$K_m (\mu M)$	$V_{max} - V_0$ [a]
Norepinephrine	17^b (5)	0.12 (1)	0.3 (5)	12.1
Nicotinamide	4.4 (7)	10 (4)	0.1 (7)	0.04

[a] Units are μmol/g tissue protein/h.
[b] Estimated from graph, taking into account enzyme purification.

which can compete with L-dopa for the enzyme. This rough calculation makes plausible the ineffectiveness of nicotinamide in contrast to L-dopa.

Similar reasoning can be used to predict the S-adenosylmethionine lowering capacity of other methyl acceptors. On this basis we postulated that guanidinoacetic acid should be a potent S-adenosylmethionine reducing agent. The V_0 for methylation of guanidino-acetic acid to creatine in rat liver is approximately 8 mg/day (estimated with an isolated perfused organ preparation) (6), whereas urinary excretion of VMA (the predominant methylated catechol metabolite) is smaller than this even in human urine (2). Without S_0 and K_m for guanidinoacetic acid methyltransferase, a computation of $V_{max} - V_0$ cannot be carried out, but the very high rate of creatine formation in the liver suggests that guanidinoacetic acid might be effective in lowering S-adenosylmethionine in that tissue. In preliminary experiments, we observed a 45% decrease in liver S-adenosylmethionine after feeding nine doses of 40 mg guanidinoacetic acid to rats over a 30day period (there was no change in the brain), but the experiments need replication before they can be regarded as conclusive. Perhaps other S-adenosylmethionine lowering agents can also be predicted in this way.

References

1. Anton, A.H., Sayre, D.F.: A study of the factors affecting the aluminum oxide-tri-hydroxyindole procedure for the analysis of catecholamines. J. Pharmacol. exp. Ther. **138**, 360 (1962)
2. Armstrong, M.D., McMillan, A., Shaw, K.N.F.: 3-Methoxy-4-hydroxy-D-mandelic acid, a urinary metabolite of norepinephrine. Biochim. biophys. Acta **25**, 422 (1957)
3. Baldessarini, R.J.: Alterations in tissue levels of S-adenosylmethionine. Biochem. Pharmacol. **15**, 741 (1966)
4. Chaudhuri, D.K., Kodicek, E.: The fluorimetric estimation of nicotinamide in biological materials. Biochem. J. **44**, 434 (1949)
5. Crout, J.R.: Inhibition of catechol-O-methyltransferase by pyrogallol in the rat. Biochem. Pharmacol. **6**, 47 (1961)
6. Gerber, G.B., Gerber, G., Koszalka, T.R., Miller, L.L.: The rate of creatine synthesis in the isolated, perfused rat liver. J. biol. Chem. **237**, 2246 (1962)
7. Salvador, R.A., Burton, R.M.: Inhibition of the methylation of nicotinamide by chlorpromazine. Biochem. Pharmacol. **14**, 1185 (1965)
8. Wurtman, R.J., Rose, C.M., Matthysse, S., Stephenson, J., Baldessarini, R.J.: L-dihydroxyphenylalanine: effect on S-adenosylmethionine in brain. Science **169**, 395 (1970)

Effects of S-Adenosyl-L-Methionine on the Behavior of Mice

M. SANSONE

The present work was prompted by recent clinical observations (4), in which antidepressant effects were claimed for S-adenosyl-L-methionine (SAMe). Our work was designed to explore the effects of SAMe on the behavior of experimental animals — namely, on mice treated with the substance either alone or associated with other drugs endowed with actions on the CNS. The animals were then exposed to experimental situations such as are usually implemented for the screening of psychoactive drugs. We chose this experimental design in an attempt to define the behavioral effects of SAMe in comparison with those of the tricyclic antidepressants.

Spontaneous Motor Activity

We explored the effects of SAMe on the spontaneous motor activity of mice by treating the animals with the substance alone or associated with amphetamine.

The apparatus, described elsewhere (12), consists of eight plexiglas boxes (floor surface 40 x 10 cm), each divided into two equal compartments communicating by way of a small aperture (3 x 3 cm). The number of crossings from one compartment to the other was recorded automatically for each mouse over a period of 60 min by the slight bascule action of the box floor. The boxes were situated in a soundproof room with dim illumination from a single bulb.

We made a preliminary experiment to determine whether SAMe alone had any effects on spontaneous motor activity in the mouse. Four groups of male mice of the heterogeneous Swiss strain (ARSAL Laboratories, Rome), 24 animals to a group were treated intraperitoneally with normal saline solution at a dosage of 10 ml/kg (control group) or with 5, 10, and 20 mg/kg of SAMe; the animals were placed in the bascule boxes 15 min after the injection.

Figure 1 shows the mean number of crossings for each group of mice over consecutive intervals of 10 min each. All groups showed an appreciable reduction of crossings between the first and the second interval, with little or no reduction of spontaneous motor activity the rest of the time. Statistical analysis of the total number of crossings in 60 min revealed no significant differences between groups (D.F. 3/92; F = 1.745; $P > 0.05$). The conclusion was that SAMe, at the dosages used in this trail, did not produce significant changes of spontaneous motor activity in mice.

Laboratory of Psychobiology and Psychopharmacology, CNR (Director: Prof. D. Bovet) - Rome, Italy.

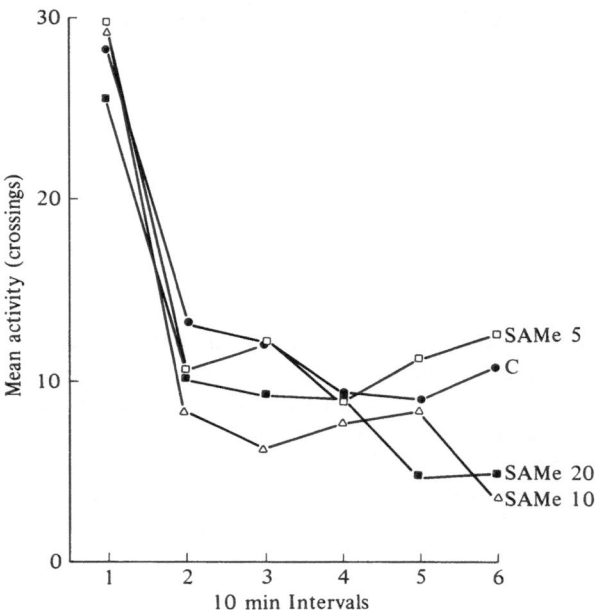

Fig. 1. Swiss strain mice. Spontaneous motor activity (number of crossings) 15 min after treatment with normal saline solution (C) or SAMe in the amounts of 5, 10, and 20 mg/kg by intraperitoneal injection. Means of values calculated in 6 consecutive intervals of 10 min each during a 60 min session

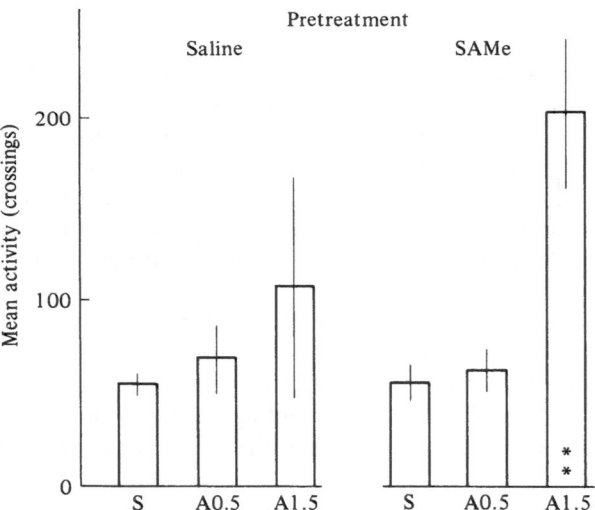

Fig. 2. Swiss strain mice. Spontaneous motor activity (mean number of crossing ± se) during a 60 min session. Pretreatment consisted of an intraperitoneal injection of normal saline solution (10 ml/kg) or SAMe (10 mg/kg), administered 15 min before the session, and followed 10 min later (i.e., 5 min before the session) by normal saline solution (S) or d-amphetamine sulfate (A) at dosages of 0.5 and 1.5 mg/kg, again by intraperitoneal injection. Animals receiving the SAMe treatment were given the same substance (10 mg/ kg daily) also for 4 consecutive days before the experiment. Statistical significance between saline and amphetamine treatments is based on Student's t-test. The double asterisk stands for $P < 0.01$

In another experiment, again with male Swiss mice, we tested the animals for possible interactions between SAMe and amphetamine; treatments and results are shown in Figure 2. The results indicate that neither SAMe nor amphetamine, administered alone, increase the number of crossings significantly. Pretreatment with SAMe (10 mg/kg) does not influence significantly the effect of amphetamine (0.5 mg/kg); conversely, the combination of SAMe with a higher dosage of amphetamine (1,5 mg/kg) results in a marked and statistically significant increase of the number of crossings. Yet it does not seem that SAMe simply potentiates the motor-stimulating action of amphetamine. Individual data (Fig. 3) show that only one of eight mice treated with amphetamine alone (1.5 mg/kg) made many crossings, whereas all the other made fewer crossings than the controls.

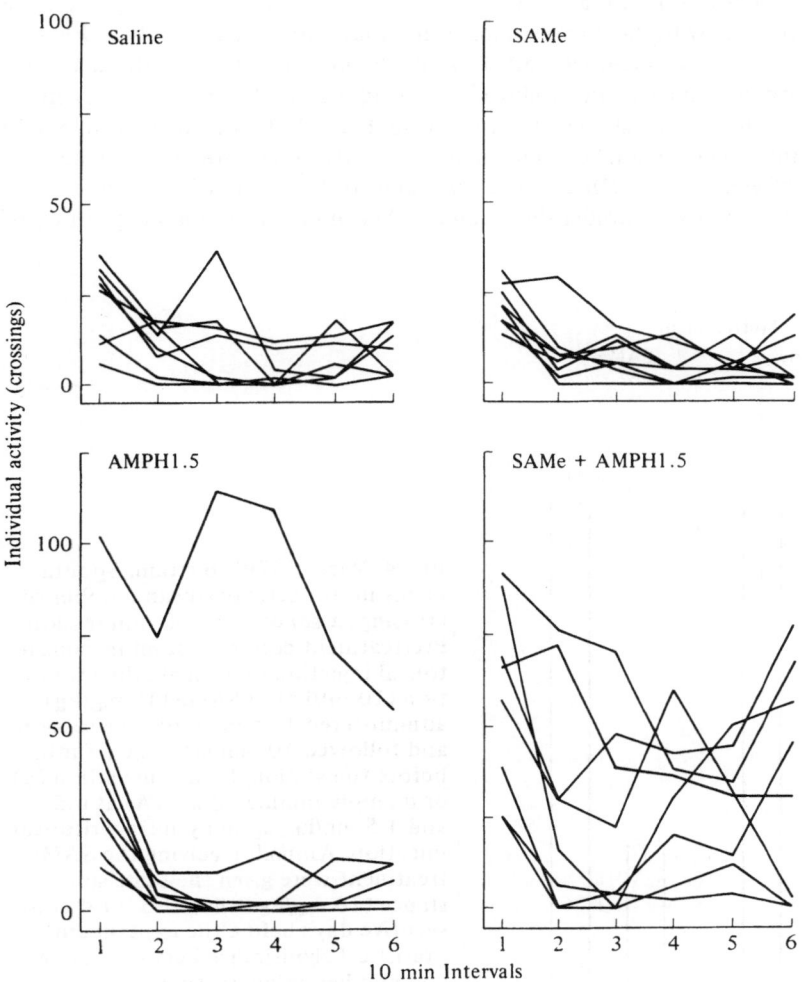

Fig. 3. Swiss strain mice. Spontaneous motor activity (crossings). Individual values for 6 consecutive intervals of 10 min each, refer to some groups of animals included in Figure 2

124

Perhaps in this strain of animals amphetamine at a dosage of 1.5 mg/kg induces some kind of stereotyped behavioral response (14) incompatible with locomotor activity; then the increased activity of mice receiving amphetamine in the stated dosage after pretreatment with SAMe might be explained at least partly as a correction of amphetamine stereotypy, or some components thereof, by SAMe.

This interpretation is apparently corroborated by the results of another experiment, in which the same dosages of SAMe and amphetamine were administered to mice from two pure strains, namely C57BL/6 and DBA/2 (ARSAL Laboratories, Rome). These two strains were chosen because they had shown a different reactivity to amphetamine in preliminary trial runs. Also, from a recent study of the effects of morphine on spontaneous motor activity (10), it emerged that under the influence of the drug, C57BL/6 mice showed a great deal of motor activity, whereas DBA/2 mice behaved sluggishly. The results of our own trials with these two strains of mice are shown respectively in Figures 4 and 5. As can be seen, treatment with SAMe alone did not modify the motor activity of either strain. Amphetamine alone produced a marked increase of motor activity in strain C57BL/6 but had practically no effect on strain DBA/2. Pretreatment with SAMe did not modify the effects of amphetamine in strain C57BL/6 but produced a striking increase of such effects in strain DBA/2. The difference of behavior of DBA/2 mice stands out more clearly if we consider the mean number of crossings separately for con-

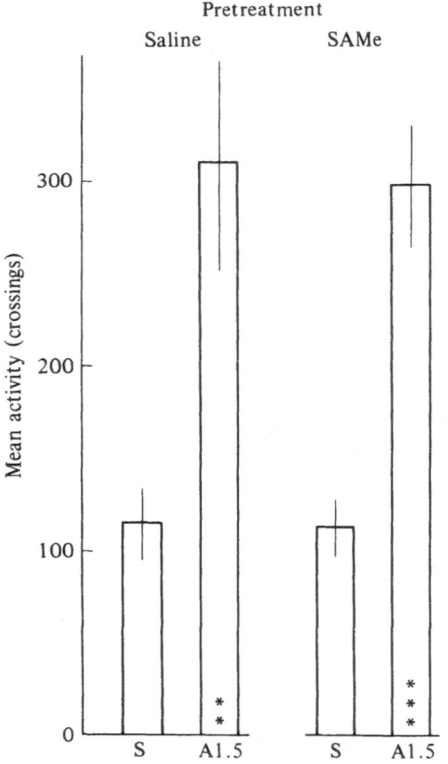

Fig. 4. Mice, C57BL/6 strain. Spontaneous motor activity (mean number of crossings ± se) during a 60 min session. Pretreatment consisted of an intraperitoneal injection of normal saline solution (10 ml/kg) or SAMe (10 mg/kg), administered 15 min before the session and followed 10 min later (i.e., 5 min before the session) by normal saline (S) or d-amphetamine sulfate (A) at 0.5 and 1.5 mg/kg, again by intraperitoneal injection. Animals receiving the SAMe treatment were given the same substance (10 mg/kg daily) also for 4 consecutive days before the experiment. Statistical significance between saline and amphetamine treatments is based on Student's t-test. The double asterisk stands for $P < 0.01$; the triple asterisk for $P < 0.001$

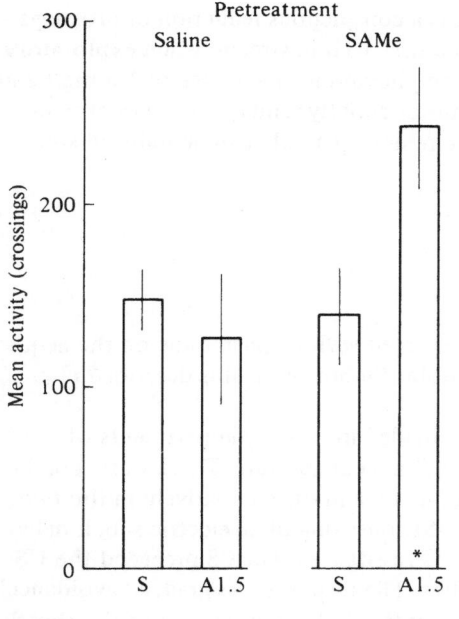

Pretreatment

Fig. 5. Mice, DBA/2 strain. Spontaneous motor activity (mean number of crossings ± se) during a 60 min session. Pretreatment was the same as in Figure 4. The asterisk stands for $P < 0.05$

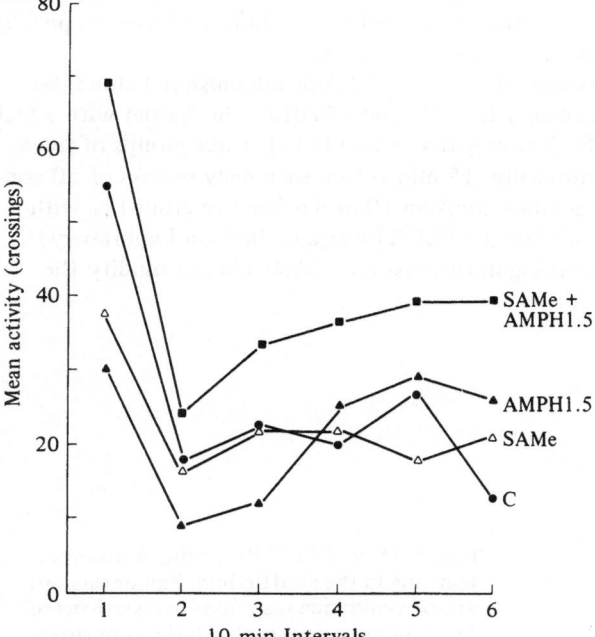

Fig. 6. Mice, DBA/2 strain. Spontaneous motor activity (crossings) in 6 consecutive intervals of 10 min each. Values refer to the groups shown in Figure 5. C = saline solution after saline solution pretreatment. SAMe = saline solution after SAMe pretreatment. AMPH 1.5 = d-amphetamine sulfate (1.5 mg/kg) after saline solution pretreatment. SAMe + AMPH 1.5 = d-amphetamine sulfate (1.5 mg/kg) after SAMe pretreatment

secutive intervals of 10 min each (Fig. 6). There is a conspicuous reduction of crossings in animals treated with amphetamine alone in the first 10 min stretch, when exploratory activity is usually greatest. Here, again, perhaps amphetamine at a dosage of 1.5 mg/kg induces a stereotyped behavior that reduces locomotor activity; and again, pretreatment with SAMe might increase locomotor activity by removing amphetamine-induced stereotypies.

Learning in Avoidance Tests

We investigated the effects of SAMe alone or associated with amphetamine on the acquisition of avoidance behavior by mice placed in the shuttle-box apparatus described elsewhere (1).

We used eight automated shuttle-boxes, each divided into two compartments of 20 x 10 cm floor space communicating through a 3 x 3 cm aperture. There was a conditional stimulus (CS) consisting of a 10 W light going on and off alternatively in the two compartments, and an unconditional stimulus (US) consisting of an electric shock delivered to the box floor (110 V through a 500,000 Ω resistance). The CS preceded the US by 5 s and covered it for 25 s. Each test run was 30 s. The response was graded "avoidance" when the mouse avoided the US by taking refuge in the dark compartment of the shuttle-box in less than 5 s of CS. Failing this, the mouse received a shock but could still escape it by shuttling during the US. Spontaneous crossings were punished.

We made a first experiment to explore the effects of SAMe administered alone. We selected two pure strains of mice, namely SEC/1Re and C57BL/6, the former with a high and the latter with a low capacity for learning avoidance (1, 11). Three groups of mice of each strain were treated intraperitoneally, 15 min before each daily session of 50 consecutive avoidance runs, with normal saline solution 10 ml/kg (control group) or with SAMe 5 mg/kg and 10 mg/kg. The mice of the SEC/1Re strain (Jackson Laboratory) achieved a high level of avoidance learning in five sessions. SAMe did not modify the

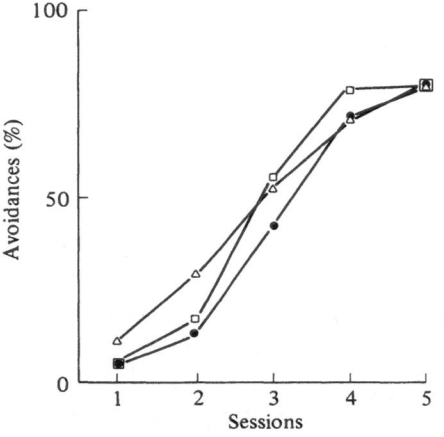

Fig. 7. Mice, SEC/1Re strain. Avoidance learning in the shuttle-box. Percentage means of avoidance responses in 5 sessions of 50 avoidance runs each. There were three groups of eight mice. Each group was treated intraperitoneally, 15 min before each daily session with either normal saline solution (black circles), or SAMe 5 mg/kg (light squares), or SAMe 10 mg/kg (light triangles)

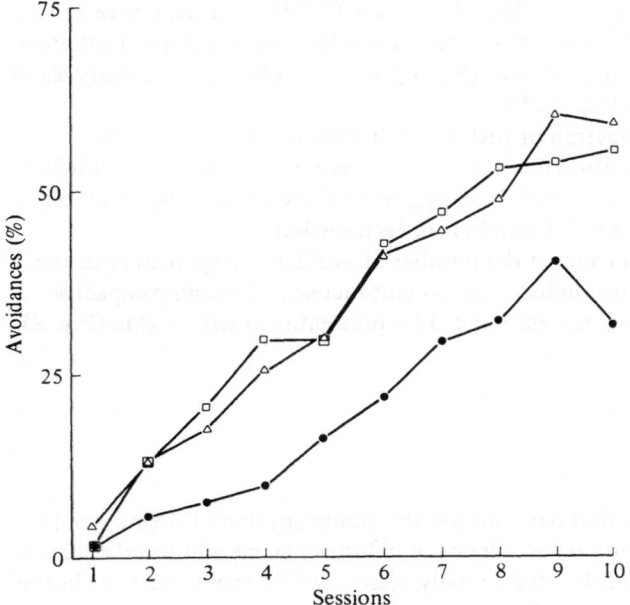

Fig. 8. Mice, C57BL/6 strain. Avoidance learning in the shuttle-box. Percentage means of avoidance responses in 10 sessions of 50 avoidance runs each. There were three groups of mice. Each group was treated intraperitoneally, 15 min before each daily session, with either normal saline solution (black circles: 8 animals), or SAMe 5mg/kg (light squares: 8 animals), or SAMe 10 mg/kg (light triangles: 16 animals)

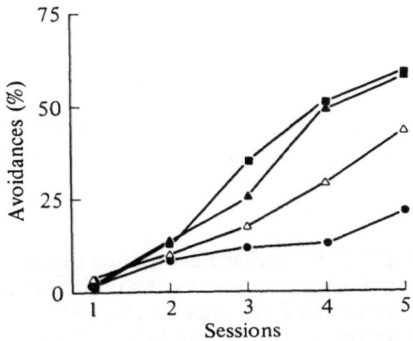

Fig. 9. Mice, C57BL/6 strain. Avoidance learning in the shuttle-box. Percentage means of avoidance responses in 5 sessions of 50 avoidance runs each. There were four groups, each of eight animals. At 15 min before each daily session, two groups were treated intraperitoneally with normal saline solution (10 ml/kg) and the other two with SAMe (10 mg/kg). After 10 min (or 5 min before each daily session) the animals were treated intraperitoneally either with normal saline solution (same dose as before) or with d-amphetamine sulfate (0.5 mg/kg). The treatment sequence was as follows: Black circles: normal saline plus normal saline; Light triangles: SAMe plus normal saline; Black triangles: normal saline plus amphetamine; Black squares: SAMe plus amphetamine

avoidance capacity in this strain (Fig. 7). The mice of the C57BL/6 strain, conversely, showed low levels of avoidance learning even after ten sessions. In this strain, both dosages of SAMe improved the learning capacity (Fig. 8), the effect being statistically significant with the 10 mg/kg dosage ($P < 0.05$).

Next we investigated the interaction of SAMe and amphetamine in mice of the C57BL/6 strain subjected to 5 sessions a day, each of 50 avoidance runs. The animals were treated with SAMe (10 mg/kg) and d-amphetamine sulfate (1 mg/kg), separately or in conjunction, respectively 15 and 5 min before each session.

Treatment with SAMe alone increased the number of avoidance responses as in the previous experiment. Amphetamine induced an evident increase of learning capacity; but the effect of amphetamine was not influenced by pretreatment with SAMe (Fig. 9).

Drug-Depressed Avoidance

This experiment was designed to find out whether the administration of SAMe would interfere with the depressant action on avoidance of chlorpromazine and tetrabenazine. Mice of the SEC/1Re strain were subjected to daily sessions of 50 avoidance runs in the shuttle-boxes. Chlorpromazine and tetrabenazine were administered intraperitoneally at several dosage levels 30 min before an avoidance test session; drug activity was assessed as described elsewhere (15) from the reduction of avoidance responses compared to a previous run without drugs (control session). In all chlorpromazine and tetrabenazine dosage groups, either normal saline solution (10 ml/kg) or SAMe (10 mg/kg) was admin-

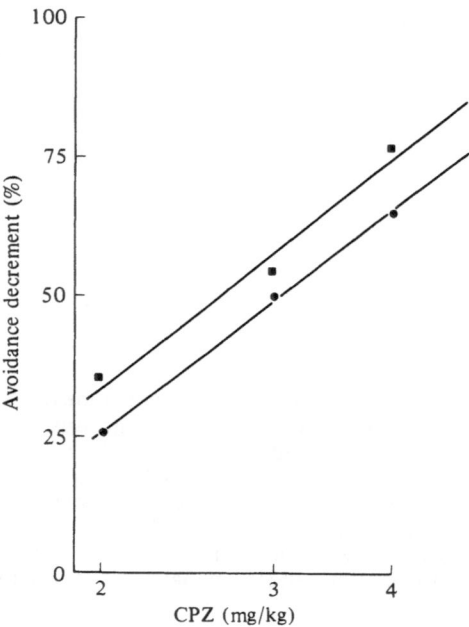

Fig. 10. Percentage avoidance decrement after treatment with chlorpromazine (CPZ) in mice of the SEC/1Re strain. Ab 30 min before the session the mice received an intraperitoneal injection of 2, 3, or 4 mg/kg of chlorpromazine; then 15 min before the session they received an intraperitoneal injection of either normal saline solution (black circles) or SAMe (10 mg/kg). Those receiving SAMe were given the same dose daily also on the 4 days preceding the test

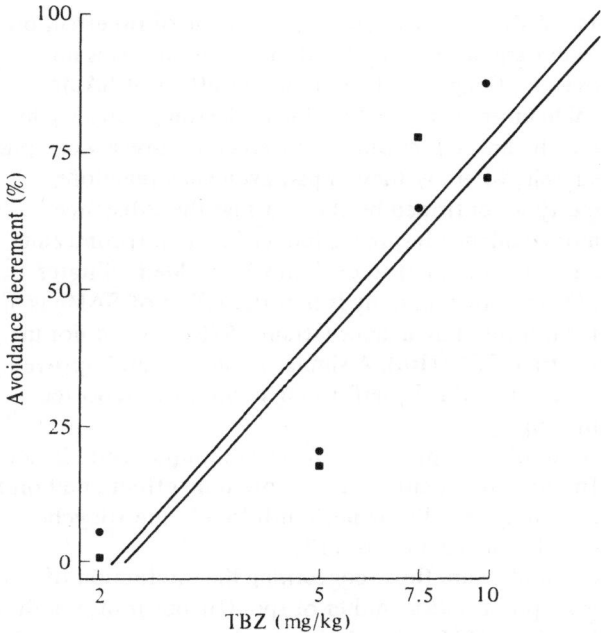

Fig. 11. Percentage avoidance decrement after treatment with tetrabenazine (TBZ) in mice of the SEC/1Re strain. Same experimental design as in Figure 10

istered intraperitoneally 15 min after the avoidance-depressing treatment. The SAMe-treated mice received similar injections of the substance in the amount of 10 mg/kg daily also for 4 days before the experiment.

The regression lines in Figure 10 (chlorpromazine) and Figure 11 (tetrabenazine) show a very definite dose-effect relationship and give no evidence of any interfering action from SAMe. The dose of chlorpromazine that induced a 50% reduction of avoidance responses (ED^{50}) was 3.06 mg/kg for mice treated with normal saline solution, and 2.66 mg/kg for those pretreated with SAMe; the respective ED^{50} values in the tetrabenazine treatment group were 5.82 and 6.27 mg/kg. Thus we may conclude that SAMe, in the dosages used in these experiments, did not modify the depression of avoidance responses induced by chlorpromazine and tetrabenazine in mice.

Discussion

There emerge from our experiments a number of interesting results about the effects of SAMe alone or in association with other treatments on the spontaneous locomotor activity and on the learning of avoidance behavior in mice. These results are admittedly not enough to warrant a detailed outline of the effects of SAMe on animal behavior, but afford some comparisons with the effects of tricyclic antidepressants.

Like the tricyclic antidepressants, SAMe administered alone has little or no effect on the behavior of "normal" animals. Spontaneous motor activity in the mouse was not modified significantly even with doses of 20 mg/kg. The only visible effect of SAMe alone, in our experiments, was a sizable improvement of avoidance learning capacity in mice of the C57BL/6 strain. The tricyclic antidepressants, contrariwise, show a weak phenothiazine-like action at high dosage levels, whereby they impair avoidance reactions (16). That SAMe is free of this property is confirmed by the fact that the substance showed no effects on the reduction of avoidance responses induced by chlorpromazine.

No ready-made answer is available at the moment to explain why SAMe facilitates the learning of avoidance behavior. One interesting thing is that this effect of SAMe is evident in mice with a low spontaneous level of avoidance (strain C57BL/6) and not in those with a high level of avoidance (strain SEC/1Re). A similar trend, i.e., an improvement of avoidance learning capacity only in animals performing poorly, was reported with the administration of stimulant drugs (3).

As for tetrabenazine, the antagonism of its depressant effects was found useful in the study of antidepressant drugs (5). In our experiments, SAMe showed no effect at all on the depression of avoidance learning induced by tetrabenazine. Truth to tell, the tricyclic antipressants are also ineffective in the tetrabenazine test (17).

The most interesting results of our studies are those concerning the interaction of SAMe and amphetamine. There was no potentiation, either of spontaneous motor activity or of avoidance learning capacity, when SAMe was administered in conjunction with a low dosage of amphetamine (0.5 mg/kg); but there was a marked increase of spontaneous activity when SAMe was associated with higher dosages of amphetamine (1.5 mg/kg). This effect, however, was seen only in the mouse strain that showed no increase of spontaneous motor activity with amphetamine alone at a dosage of 1.5 mg/kg. This pattern suggests that SAMe does not simply potentiate the effects of amphetamine on behavior but probably acts by way of counteracting certain stereotyped responses induced by amphetamine. Imipramine, conversely, potentiates amphetamine-induced stereotypy (7).

The inhibition of amphetamine-induced stereotypies has been reported with neuroleptic drugs, allegedly as a result of dopaminergic antagonism (6, 8, 9, 13). At any rate, perhaps a better explanation of the SAMe-amphetamine interaction may be found in terms of SAMe-dependent transmethylation processes (2). In addition to a possible methylation of amphetamine, which might change its effects on behavior, we must take into account the possible chemical transformations of the catechols that are liberated by amphetamine administration.

Summing up: while further experiment work is certainly needed for a more thorough characterization of SAMe, what we have learned so far seems enough to rule out any resemblance between the mechanisms of antidepressant action of this substance (4) and those of imipraminelike drugs.

References

1. Bovet, D., Bovet-Nitti, F., Oliverio, A.: Genetic aspects of learning and memory in mice. Science 163, 139 (1968)

2. Cohn, C.K., Vesell, E.S., Axelrod, J.: Studies of a methionine-activating enzyme. Biochem. Pharmacol. **21**, 803 (1972)
3. Essman, W.B.: Drug effects and learning and memory processes. Advanc. Pharmacol. Chemother. **9**, 241 (1971)
4. Fazio, C., Andreoli, V., Agnoli, A., Casacchia, M., Cerbo, R.: Effetti terapeutici e meccanismo d'azione della S-adenosil-L-metionina (SAMe) nelle sindromi depressive. Minerva med. (Torino) **64**, 1515 (1973)
5. Gyermek, L.: The pharmacology of imipramine and related antidepressants. Int. Rev. Neurobiol. **9**, 143 (1966)
6. Klingenstein, R.J., Wallach, M.B., Gershon, S.: A comparison of pimozide and thioridazine as antagonists of amphetamine-induced stereotyped behavior in dogs. Arch. int. Pharmacodyn. **203**, 67 (1973)
7. Kulkarni, S.K., Dandiya, P.C.: On the mechanism of potentiation of amphetamine-induced stereotype behavior by imipramine. Psychopharmacologia **27**, 367 (1972)
8. Lal, S., Sourkes, T.L.: Potentiation and inhibition of the amphetamine stereotype in rats by neuroleptics and other agents. Arch. int. Pharmacodyn. **199**, 289 (1972)
9. Lemberger, L., Witt, E.D., Davis, J.M., Kopin, I.J.: The effects of haloperidol and chlorpromazine on amphetamine metabolism and amphetamine stereotype behavior in the rat. J. Pharmacol. exp. Ther. **174**, 428 (1970)
10. Oliverio, A., Castellano, C.: Genotype dependent sensitivity and tolerance to morphine and heroine: dissociation between opiate-induced running-fit and analgesia in the mouse. Psychopharmacologia. In press (1977)
11. Oliverio, A., Castellano, C., Messeri, P.: A genetic analysis of avoidance, maze, and wheel-running behaviors in the mouse. J. comp. physiol. Phsychol. **79**, 459 (1972)
12. Oliverio, A., Castellano, C., Messeri, P.: Genotype-dependent effects of septal lesions on different types of learning in the mouse. J. comp. physiol. Psychol. **82**, 240 (1973)
13. Randrup, A., Munkvad, I.: Special antagonism of amphetamine-induced abnormal behavior. Inhibition of stereotyped activity with increase of some normal activities. Psychopharmacologia **7**, 416 (1965)
14. Randrup, A., Munkvad, I.: Stereotyped activities produced by amphetamine in several animal species and man. Psychopharmacologia **11**, 300 (1967)
15. Sansone, M., Messeri, P.: Strain differences on the effects of chlordiazepoxide and chlorpromazine in avoidance behavior of mice. Pharmacol. Res. Commun. **6**, 179 (1974)
16. Sigg, E.B.: Tricyclic thymoleptic agents and some newer antidepressants. In: Psychopharmacology: A Review of Progress 1957-1967. Efron, D.H. (ed.) Public Health Service Publication, No. 1836, Washington, 1968, pp. 655-669
17. Voith, K., Herr, F.: The effect of various antidepressant drugs upon the tetrabenazine-suppressed conditioned avoidance response in rats. Psychopharmacologia **20**, 253 (1971)

Effects of Exogenous L-Dopa on the Metabolism of Methionine and S-Adenosylmethionine in the Brain

R.J. WURTMAN[1] and L.A. ORDONEZ[2]

The catecholamines dopamine, norepinephrine, and epinephrine are metabolized by oxidative deamination and by O-methylation. This latter pathway is catalyzed by the enzyme catechol-O-methyl-transferase (1); it utilizes S-adenosyl-methionine (SAMe) as the methyl donor (Fig. 1), and its products are, for the most part, methylated on the metahydroxyl group. The catecholamines themselves are formed within only a few specific tissues in the body (21). These include a relatively small number of brain neurons, postganglionic sympathetic neurons, adrenomedullary chromaffin cells, and interneurons within sympathetic ganglia. The catechol-amines can be O-methylated within their cells of origin, within synapses, or, after secretion into the blood stream, within the liver, kidneys, and other organs.

When humans or experimental animals are given the catechol-amine precursor L-dihydroxyphenylalanine (L-dopa), much of this exogenous amino acid is also O-methylated prior to its excretion (6, 22). The dopa is initially taken up within virtually all cells, probably by the same uptake systems as those that mediate the uptake of tyrosine and phenylalanine (19). Thereafter cells that contain the enzyme aromatic L-amino acid decarboxy-lase (AAAD: "dopa decarboxylase") convert it to the catechol-amine dopamine. This enzyme is probably present within most cells in the brain and other tissues (12, 19). Noradrenergic neurons can further transform it to norepinephrine. Both dopa and the catecholamines formed from it are excellent substrates for catechol-O-methyltransferase. Because the doses of dopa generally used tend to be relatively large, correspondingly large quantities of SAMe are utilized in its transmethylation. As described below, this causes the levels of SAMe in brain and ultimately some other tissues to decline. Brain methionine levels tend not to fall, even after major decreases in brain SAMe. This observation has led to the discovery that brain contains a folate-dependent enzyme system which can catalyze the regeneration of methionine from the homocysteine formed after SAMe has given up its methyl group.

[1] Laboratory of Neuroendocrine Regulation, Department of Nutrition and Food Science, Massachusetts Institute of Technology, Cambridge, Massachusetts, USA.
[2] Catedra de Fisiopatologia, Instituto de Medicina Experimental, Universidad Central de Venezuela, Caracas, Venezuela.

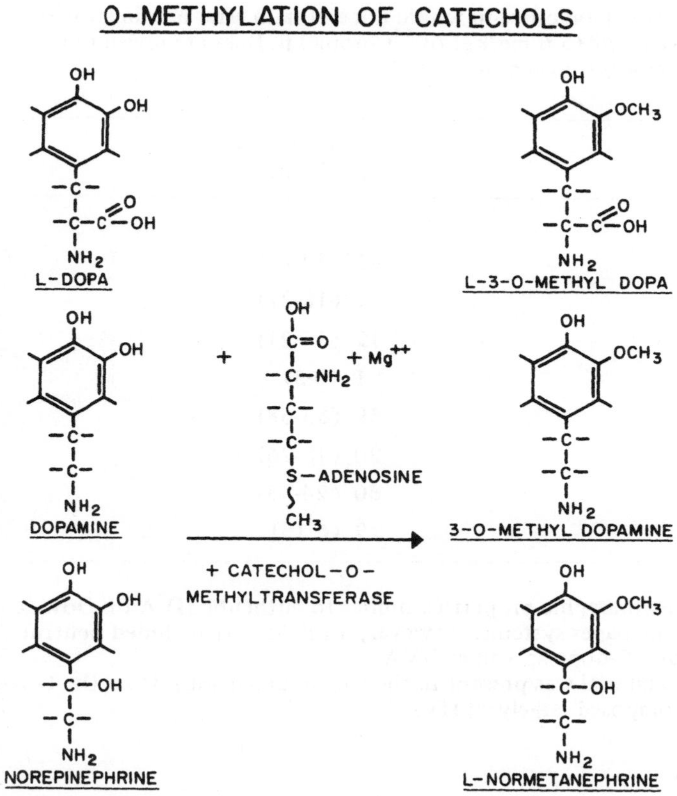

O-METHYLATION OF CATECHOLS

Fig. 1. Enzymatic O-methylation of various catechols by catechol-O-methyltransferase. A methyl group is transferred from S-adenosylmethionine to the metahydroxy position of dopa, dopamine, norepinephrine, or epinephrine

O-Methylation of Exogenous Dopa in the Whole Mouse

Bartholini and Pletscher (3) noted the presence of relatively large quantities of ^{14}C-3-O-methyl dopa in brains of rats given ^{14}C-dopa 60 min earlier. The accumulation of this compound was attributed to the fact that it is a poor substrate of AAAD (8) and thus, once formed, its catabolism would probably be much slower than that of dopa itself. In order to examine the fate of exogenous L-dopa, we identified and measured its various radioactive metabolites in homogenates of white mice given ^{14}C-dopa intraperitoneally (22). A very major fraction of the radioactivity present as early as 20 min after ^{14}C-dopa administration was found to consist of O-methylated metabolites, of which the major constituent was methoxydopa (3-O-methyl dopa) (Table 1). This observation suggested that very large amounts of S-adenosylmethionine (and thus of methionine) would be

Table 1. Identity of radioactive metabolites in whole mice 20 min after administration of ^{14}C-dopa. Six mice received 5 μc (5-6 mg/kg) of ^{14}C-dopa i.p. Data are presented as average and range (in parentheses) for each metabolite

	Percentage of metabolite
	%
Alumina eluate	41 (32-47)
Dopa	21 (18-27)
Catecholamines	19 (15-21)
Unidentified compounds	1 (0-3)
Alumina effluent	59 (53-68)
"HVA"[a]	20 (18-26)
Methoxydopa	30 (24-33)
Unidentified compounds[b]	9 (6-15)

[a] This material had chromatographic properties similar to authentic HVA on Dowex columns and on two ascending paper systems. However, it might have included neutral O-methylated metabolites of ^{14}C-dopa as well as HVA.
[b] About one-half of this material was present in the first aqueous wash from the Dowex column and was probably composed largely of HVA.
(Reprinted from (22)).

required to metabolize the doses of dopa (50-100 mg/kg) routinely taken by patients treated for Parkinson's disease. Support for this hypothesis was provided by the finding of Calne (6) that considerably more than half of the total urinary metabolites of dopa in men receiving the amino acid chronically are O-methylated.

Depletion of Brain S-Adenosylmethionine by Exogenous Dopa

To determine whether the O-methylation of exogenous L-dopa affected the concentration of SAMe in the brain, the methyl donor was assayed in groups of rats that had received 0, 10, 30 or 100 mg of the catechol amino acid per kilogram in the middle of the daily dark period and were killed 45 min later.

(Previous studies had shown that brain SAMe levels vary diurnally, reaching their peak at this time of day.) Treatment with 30 mg/kg of L-dopa was associated with a 36% reduction in brain SAMe (Table 2); treatment with 100 mg/kg caused a 67% reduction ($P < 0.01$) (23). As little as 10 mg/kg of L-dopa caused a 51% decrease in concentration of SAMe in the adrenal medulla ($P < 0.01$); in contrast, the content of SAMe in the liver was not significantly altered 45 min after the injection of as much as 100 mg/kg of L-dopa. In other experiments, it was noted that the decrease in SAMe concentration in brains of rats rats receiving 100 mg/kg of L-dopa was at least as pronounced in the middle

Table. 2. Relation between dose of L-dopa and extent of depletion of SAMe content in tissue. Groups of five rats received L-dopa i.p. and were killed 45 min later. Data are presented as mean concentration of SAMe (μg per g of wet tissue) ± SE of the mean

Tissue	L-dopa dose (mg/kg)			
	0	10	30	100
Brain	16.8 ± 0.6	16.0 ± 1.1	10.7 ± 0.3	5.5 ± 0.4*
Adrenal	39.4 ± 1.8	19.4 ± 3.9+	15.1 ± 4.9+	14.1 ± 3.6*
Liver	56.8 ± 3.5	65.9 ± 2.0	71.2 ± 5.3	61.4 ± 5.4

* P < 0.001 differs from control group.   + P < 0.01 differs from control group.

(Reprint from (23)).

of the daily light period (that is, when SAMe is normally lowest) as in the middle of the dark period. Brain SAMe concentrations were found to have returned to normal with 6 h of dopa administration.

The marked depletion in brain SAMe observed soon after dopa administration raised the possibility that insufficinet quantities of the methyl donor might be available to allow the transmethylation of other methyl acceptors (e.g., the catecholamines dopamine and norepiniphrine) to proceed at normal rates. To examine this possibility, the metabolism of brain ^{3}H-norepinephrine, labelled by injecting ^{3}H-norepinephrine intracisternally, was examined in animals given dopa 5 min later. A typical study is illustrated in Figure 2; in this study animals had received daily injections of L-dopa (100 mg/kg) for 10 days prior to, as well as 5 min after the intracisternal ^{3}H-norepinephrine. Dopa administration causes a transient major depletion of brain SAMe; the nadir (20% of control values) was attained 1 h after rats received the catechol amino acid. At this time, the proportion of brain ^{3}H-norepinephrine metabolites containing O-methyl groups was decreased by about half, while the ^{3}H-deaminated, non-O-methylated metabolites were correspondingly increased (7). Dopa administration decreased the accumulation of O-methylated norepinephrine metabolites in all brain regions examined, but especially in the telencephalon, hypothalamus, and brain stem (18). This suppression of O-methylation might be expected to potentiate the physiologic effects of norepinephrine molecules released into central synapses. Dopa administration may also enhance central noradrenergic transmission by increasing brain norepinephrine levels (7) and by accelerating the turnover of this monoamine (7, 18).

Effects of Dopa on S-Adenosylmethionine and Methionine

Levels in Various Tissues

The studies described above raised the possibility that the methylation of large amounts of exogenous L-dopa might ultimately decrease the pools of free methionine in the brain and other tissues. If dietary methionine provided the only source of methyl groups for SAMe synthesis, a 70 kg human taking 6-10 of L-dopa should ultimately become methionine-deficient. However, it had been demonstrated that, even after prolonged administration of L-dopa, the percentage of the catechol amino acid that is excreted in urine as methylated products remains very high (6). To examine the effects of dopa-induced SAMe depletion on methionine concentrations in blood and in various tissues, we measured these compounds in specimens obtained from rats killed at various intervals after receiving one or more doses of L-dopa (100 mg/kg). A tRNA loading assay was developed which allowed us to assay small amounts of methionine (13).

Rats received one, two, three, or four injections of L-dopa at 45-min intervals, and were killed 45 min after the last dose. SAMe concentrations were assayed in brain, liver, kidney, and muscle. Control rats either received no injections or were killed 45 min after the last of four injections of the vehicle alone. SAMe concentrations (μg/g ± SE) in tissues from control animals were: brain, 10.0 ± 0.8; liver, 38.7 ± 3.1; kidney, 18.7 ± 1.9; muscle, 52.2 ± 5.6. A single L-dopa injection markedly decreased the SAMe concentrations of brain and kidney, but not of liver or muscle (Fig. 3) (13). Additional L-dopa

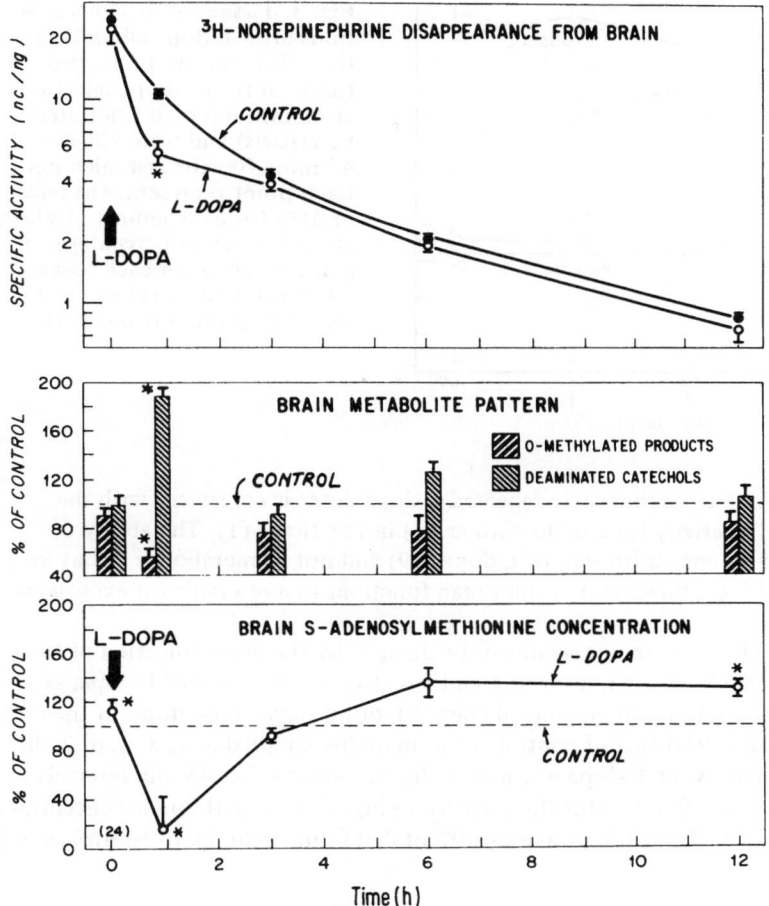

Fig. 2. ^{3}H-norepinephrine metabolism and SAMe concentration after L-dopa administration. *Top panel:* Disappearance of ^{3}H-norepinephrine. *Middle Panel:* Concentration of tritiated metabolites of ^{3}H-norepinephrine in animals given L-dopa, expressed as a percentage of the concentrations in the corresponding control group. The O-methylated products include O-methylated amine metabolites and O-methylated, deaminated compounds. *Bottom panel:* Brain SAMe concentrations in L-dopa-treated rats expressed as a percentage of the levels found in the corresponding control group. The "zero-time" animals last received L-dopa 24 h prior to intracisternal ^{3}H-norepinephrine and then again 5 min after the ^{3}H-norepinephrine. Statistically significant changes are indicated by an asterisk (Reprint from (7)).

doses did not further decrease brain SAMe; the second dose did further deplete kidney SAMe concentrations, but subsequent doses were without effect on this organ. Hepatic SAMe fell significantly (to 70% of control levels) only after three L-dopa injections. It declined further (to 53% of control values) after four injections. There was no change in SAMe concentration of muscle at any of the times studied; after four doses of L-dopa, muscle SAMe was still 91% of the amount present in tissue from control animals. This

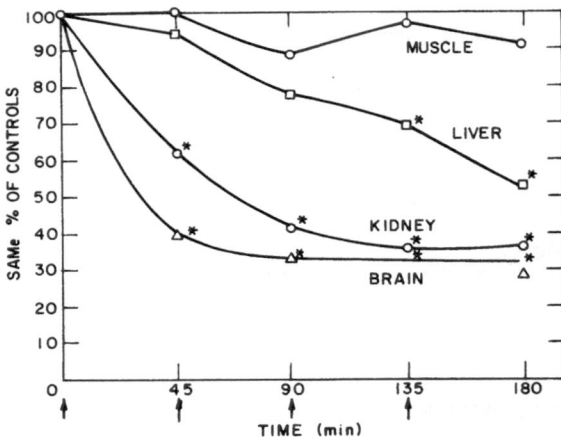

Fig. 3. Tissue SAMe concentrations after L-dopa administration. Rats received one, two, three, or four L-dopa injections at 45 min intervals (indicated by arrows) and were killed 45 min after the last injection. Each point represents the mean of data from six animals. Data are expressed as percentages of control values for each tissue. * P < 0.001 differs from controls. (Reprinted from (13)).

failure of muscle SAMe to decline after repeated L-dopa doses is consistent with the known lack of COMT activity (i.e., by in vitro assay) in this tissue (1). The ability of muscle to take up large concentrations of L-dopa (19) and not to metabolize it, but to release it unchanged (16), suggests that this organ functions as a reservoir for exogenous L-dopa.

Brain methionine levels were not significantly changed by the acute injection of L-dopa, (35.6 ± 1.7 and 37.1 ± 2.8 nmol/g at 1 and 3 h after a single dose of L-dopa, as compared with 35.6 ± 1.1 nmol/g in control animals), or by single daily doses of the amino acid for 20 days (91-116% of control values in brains sampled at 1, 3, 6, or 24 h after the last injection). When L-dopa was repeatedly administered at 45 min intervals, brain methionine began to fall after the third injection (Fig. 4). The methionine concentration 45 min after the fourth injection was only 69% of that found in brains of control animals.

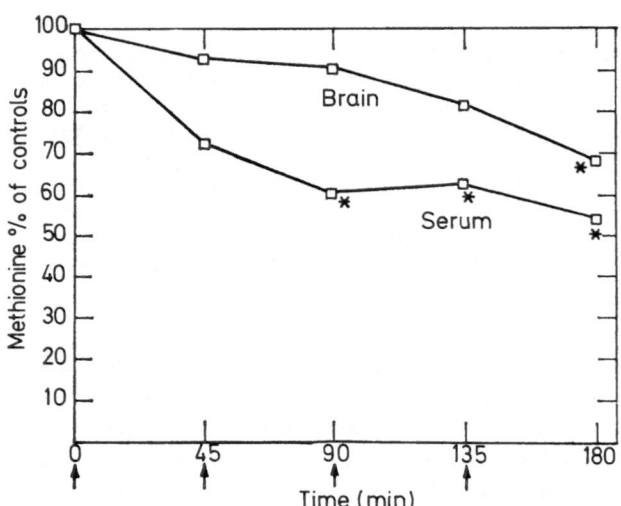

Fig. 4. Brain and serum methionine concentrations after L-dopa administration. Rats received one, two, three, or four L-dopa injections at 45 min intervals (indicated by arrows) and were killed 45 min after the last injection. Each point represents the mean of data from six animals. Data are expressed as percentages of control values for each tissue. * P < 0.001 differs from controls. (Reprinted from (13)).

Serum methionine concentrations were unchanged 1 h after a single dose of L-dopa, or after the last of ten daily doses. These concentrations were 79.7 ± 5.0 nmol/ml for control rats, and 77.0 ± 4.6 or 64.4 ± 5.8 nmol/ml for animals receiving L-dopa acutely or chronically Repeated injections of dopa at 45 min intervals did deplete serum methionine by 47% (Fig. 2).

Brain SAMe, brain methionine and serum methionine thus do not exhibit parallel responses to exogenous L-dopa. A single dose of L-dopa depresses brain SAMe (Fig. 3), but does not reduce brain or serum methionine concentrations (Fig. 4). Serum methionine content falls significantly after two or three L-dopa doses given at 45 min intervals (Fig. 4); however, brain methionine does not fall until the animal has received four such injections.

The disparity between the marked changes in brain SAMe and the absence of changes in brain methionine after a single dose of L-dopa suggested that a mechanism might be operating within brain cells for regulating their concentration of methionine. This hypothesis was supported by the observation that brain methionine levels remained within their normal range even after L-dopa treatment had lowered serum methionine (Fig. 4). (As described below, subsequent studies demonstrated that brain has the ability to regenerate methionine from homocysteine. This apparently allows it to couple methionine synthesis to SAMe utilization).

Synthesis of Methionine in Rat Brain by Folate-Dependent Enzymes

The failure of brain methionine concentrations to fall even when unusually large amounts of methionine must have been utilized to from SAMe suggested that brain might be capable of regenerating methionine from homocysteine. Early studies by Langer (10) were interpreted as showing that brain, unlike liver and kidney, was incapable of synthesizing the methyl group of methionine. However, subsequent reports have identified within rodent or bovine brain one or more of the three enzymes necessary for this process [N^5-methyl tetrahydrofolate-homocysteine (cobalamin) methyltransferase; (B_{12}-transmethylase; (9); N^5-methyltetrahydrofolate-NAD oxidoreductase (methylene reductase, EC 1.1.1.68; (5); and L-serine-tetrahydrofolate-5, 10-hydroxymethyltransferase (serine transhydroxymethylase, EC 2.1.2.1.; (4, 5)/.

Using tissues from Sprague-Dawley rats, we found that brain does contain all three of these enzymes, and can regenerate methionine from the homocysteine formed after-S-adenosylmethionine has transferred a methyl group to dopa or to other methyl acceptors (14). Brain enzyme levels were lower than those of liver or kidney. Dopa administration was found to suppress the B_{12}-transmethylase activity of dialyzed brain extracts, probably by depriving this enzyme of catalytic quantities of S-adenosylmethionine (Table 3). Since the brain is also able to concentrate N^5-methyl tetrahydrofolic acid from the circulation (11), it probably can obtain this compound for B_{12}-dependent transmethylation from two sources i.e., endogenous synthesis from glucose (via serine) (14), and uptake.

140

Table 3. The SAMe-independent B_{12}-transmethylase activity of rat tissue extracts: Effect of L-dopa treatment

| | Percentage of full activity | | |
	Brain	Liver	Kidney
Control	20.3	4.7	16.7
Acute dopa (100 mg/kg)	10.2*	4.5	---
Chronic dopa (100 mg/kg)	12.9+	3.8	15.5
Acute dopa (250 mg/kg)	7.7+	4.7	15.5
Chronic dopa (250 mg/kg)	8.2+‡	6.2	16.5

Groups of six animals weighing 100 g were killed 1 h after receiving one or the last of 10 daily injections of L-dopa i.p. or its diluent (0.05N-HCl). Dialyzed tissue extracts from each animal were assayed by duplicate determinations in the complete system (+ SAMe) or without added SAMe (SAMe-independent activity). Data represents the percent of full activity remaining in the SAMe system.
* $P < 0.01$ differs from control.
+ $P < 0.001$ differs from control.
‡ $P < 0.05$ differs from chronic dopa (100 mg/kg).
(Reprinted from 14).

Effect of Folate Deficiency on the Maintenance of Brain Methionine Concentration After L-Dopa Administration

The above studies suggested that the failure of brain methionine to decline following doses of L-dopa which deplete it of SAMe and which lower serum methionine, resulted from the brain's ability to regenerate methionine from homocysteine. Since the methionine-regenerating enzymes utilize folic acid as the carrier of the methyl group, it was possible to test this hypothesis by determining whether L-dopa affected brain methionine in folic acid-dificient rats. Animals were given a commercial folate-deficient diet for 8½ weeks; control rats received either this diet supplemented with folic acid (5 mg/kg diet) or a standard rat chow. Body and brain weights measured at the end of this period did not differ significantly between the folic-deficient and folic-supplemented groups; serun folic acid concentrations were 18.7 ng/ml in the former group and 200 ng/ml in the latter (15).

Folic deficiency alone significantly lowered brain SAMe (by about 25%), but had no effect on brain methionine (Fig. 5). Dopa administration lowered brain SAMe by a considerably greater percentage in folate-deficient than in control rats; it also caused methionine concentrations to fall significantly in the brains and sera of folate-deficient animals (Fig. 5).

These observations were interpreted as confirming the hypothesis that the brain's capacity to regenerate methionine by a folate-dependent enzymatic pathway buffers brain SAMe against the increased demand for methyl groups caused by L-dopa administration.

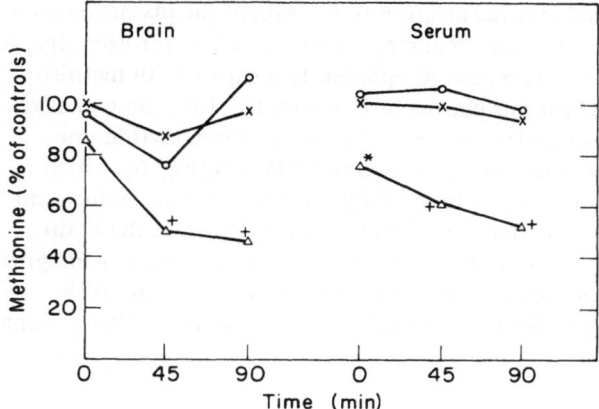

Fig. 5. Methionine concentrations of folic-deficient rats at different times after administration of L-dopa. Methionine was measured in brain and serum of groups of six animals maintained on folic-deficient (△), folic-supplemented (o), or control (x) diets at 0, 45, or 90 min after a single administration of L-dopa (100 mg/kg). The results are expressed as percentages of the control group at 0 min. Absolute values were: brain, 35.6 ± 1.7 nmol/g; serum, 79.7 ± 5.0 nmol/ml. * $P < 0.02$ differs from control groups; + $P < 0.01$ differs from control groups. (Reprinted from (15)).

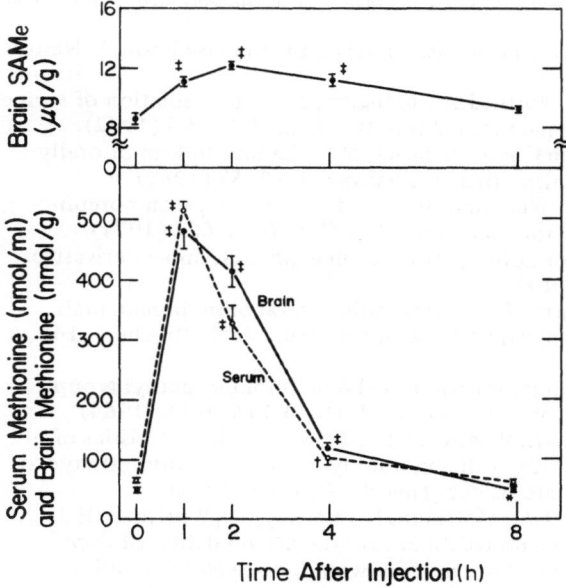

Fig. 6. Brain SAMe and serum and brain methionine concentrations at various times after injection of methionine (100 mg/kg, i.p.). Injections were timed so that all animals would be killed at the same time of day, in order to avoid differences caused by diurnal variations. Significance of differences was determined with respect to zero-time control values * $P < 0.02$ + $P < 0.01$ + $P < 0.001$

They also raise the possibility that chronic dopa administration might increase the daily requirements for folic acid (and, possibly, B_{12}) in parkinsonian patients.

The brain is also able to obtain methionine from the circulation. At any given time, brain methionine levels in control (i.e., not dopa-treated) animals are well correlated with the ratio of the plasma methionine concentration to the sum of the plasma concentrations of the other neutral amino acids (e.g., leucine, isoleucine, valine, tyrosine, tryptophan, phenylalanine (20). (These latter compounds apparently compete with methionine for transport into the brain.) When this plasma ratio is elevated, for example, after methionine is injected, or after consumption of certain foods (20) brain methionine levels rise. This, in turn, elevates the brain concentration of SAMe (2) (Fig. 6), which indicates that the methionine-activating enzyme is not fully saturated with its amino acid aubstrate in vivo (17). It seems probable that the plasma normally provides the brain with most of the methionine that it needs, and that the methionine-regenerating enzyme system described above becomes important only when unusually large amount of S-adenosylmethionine have been utilized for transmethylation reactions (e.g., after L-dopa administration).

References

1. Axelrod, J., Tomchick, R.: Enzymatic O-methylation of epinephrine and other catechols. J. biol. Chem. **233**, 702 (1958)
2. Baldessarini, R.J.: Alterations in tissue levels of S-adenosylmethionine. Biochem. Pharmacol. **15**, 741 (1966)
3. Bartholini, G., Pletscher, A.: Cerebral accumulation and metabolism of C^{14}-dopa after selective inhibition of peripheral decarboxylase. J. Pharmacol. exp. Ther. **161** 14 (1968)
4. Bridgers, W.F.: Serine transhydroxymethylase in developing mouse brain. J. Neurochem. **15**, 1325 (1968)
5. Broderich, D.S., Candland, K.L., North, J.A., Mangum, J.H.: The isolation of serine transhydroxymethylase from bovine brain. Arch. Biochem. **148**, 196 (1972)
6. Calne, D.B., Karoum, F., Ruthven C.R.J., Sandler, M.: The metabolism of orally administered L-dopa in Parkinsonism. Brit. J. Pharmacol. **37**, 57 (1969)
7. Chalmers, J.P., Baldessarini, R.J., Wurtman, R.J.: Effects of L-dopa on norepinephrine metabolism in the brain. Proc. nat. Acad. Sci. (Wash) **68**, 662 (1971)
8. Ferrini, R., Glässer, A.: In vitro decarboxylation of new phenylalanine derivatives. Biochem. Pharmacol. **13**, 798 (1964)
9. Finkelstein, J.D., Kyle, W.E., Harris, B.J.: Methionine metabolism in mammals. Regulation of homocysteine methyltransferase in rat tissue. Arch. Biochem. **146**, 84 (1971)
10. Langer, B.W.: Organ and intracellular location of the methionine methyl group synthesizing system of the rat. Proc. Soc. exp. Biol. (Ny.) **115**, 1088 (1964)
11. Levitt, M., Nixon, P.F., Pincus, J.H., Bertino, J.R.: Transport characteristics of folates in cerebrospinal fluid; a study utilizing doubly labeled 5-methyltetrahydrofolate and 5-formyltetrahydrofolate. J. clin. Invest. **50**, 1301 (1971)
12. Lytle, L.D., Hurko, O., Romero, J.A., Cottman, K., Leehey, D., Wurtman, R.J.: The effects of 6-hydroxydopamine pretreatment on the accumulation of dopa and dopamine in brain and peripheral organs following L-dopa administration. J. Neural Transm. **33**, 63 (1972)
13. Ordonez, L.A., Wurtman, R.J.: Methylation of exogenous 3, 4-dihydroxyphenylalanine: Effects on methyl group metabolism. Biochem. Pharmacol. **22**, 134 (1973a)
14. Ordonez, L.A., Wurtman, R.J.: Enzymes catalyzing the de novo synthesis of methyl groups in the brain and other tissues of the rat. J. Neurochem. **21**, 1447 (1973b)
15. Ordonez, L.A., Wurtman, R.J.: Folic acid deficiency and methyl group metabolism in rat brain: Effects of L-dopa. Arch. Biochem. **160**, 372 (1974)

16. Ordonez, L.A., Ambrus, M., Boyson, S., Goodman, M.N., Ruderman, N.B., Wurtman, R.J.: Skeletal muscle: Reservoir for exogenously administered L-dopa. J. Pharmacol., exp. Ther. **190**, 187 (1974)
17. Pan, F., Tarver, H.: Comparative studies on methionine, selenomethionine, and their ethyl analogues as substrates for methionine adenosyltransferase from rat liver. Arch. Biochem. **119**, 429 (1967)
18. Romero, J.A., Chalmers, J.P., Cottman, K., Lytle, L.D., Wurtman, R.J.: Regional effects of L-dihydroxyphenylalanine (L-dopa) on norepinephrine metabolism in rat brain. J. Pharmacol. exp. Ther. **180**, 277 (1972)
19. Romero, J.A., Lytle, L.D., Ordonez, L.A., Wurtman, R.J.: Effects of L-dopa administration on the concentrations of dopa, dopamine, and norepinephrine in various rat tissues. J. Pharmacol. exp. Ther. **184**, 67 (1973)
20. Rubin, R.A., Ordonez, L.A., Wurtman, R.J.: Physiological dependence of brain methionine and SAMe concentrations on the serum amino acid pattern. J. Neurochem. **23**, 227 (1974)
21. Wurtman, R.J.: Catecholamines. Boston: Little, Brown and Co., 1966
22. Wurtman, R.J., Chou, C., Rose, C.: The fate of C^{14}-dihydroxyphenylalanine in the whole mouse. J. Pharmacol. exp. Ther. **174**, 351 (1970a)
23. Wurtman, R.J.: Rose, C.M., Matthysse, S., Stephenson, J., Baldessarini R.J.: L-dihydroxyphenylalanine: Effect on S-adenosylmethionine in brain. Science **169**, 395 (1970b)

Human Blood Kinetics of S-Adenosyl-L-Methionine (SAMe)

V.M. ANDREOLI[1], F. MAFFEI[1], and G.C. TONON[2]

S-adenosyl-L-methionine (SAMe), a physiologic substance present in all living organisms, is very unstable at room temperature; however, BioResearch Laboratories of Milan have recently succeeded in stabilizing this molecule without changing its biological activity in any way. The solution of this problem made it possible to proceed from merely biochemical investigations of SAMe to pharmacologic work in animals and clinical trials in man.

Considering the importance of transmethylation processes in human physiology, it seems that we now have the opportunity of doing some direct testing of pathogenetic hypotheses that attribute to this natural donor of methyl radicals an important role in some cases of human pathology: a deficiency or malfunction of transmethylation processes being advocated not only in the CNS (3, 7) but also, for instance, in certain liver diseases (4).

The intraperitoneal LD_{50} of SAMe in the rat is somewhere between 2000 and 2500 mg/kg (5). The acute toxicity of the substance, however, cannot be assessed by other routes of administration because it becomes impossible to tell whether mortality should be imputed to the specific toxicity of SAMe or to the inordinately large volume of solution needed for administration. In the rat, at any rate, dosages up to 2000 mg/kg have been administered both intravenously and intramuscularly without causing any deaths. Thus the estimated LD_{50} value would represent something like 100 g for a man of medium body build.

We have explored the human blood kinetics of the SAMe molecule by injecting the substance intravenously in to two healthy male volunteers, aged 25 and 35, previously found to be free of liver, kidney, or cardiovascular pathology. The dose was 0.5 mg/kg, calculated from recommended clinical dosages reported to be therapeutically active (2).

We obtained a preliminary blood sample just before administration; then we drew a number of further samples, 7 ml each, at various intervals up to 180 min, storing each sample in a separate test tube with some EDTA disodium salt. Finally we assayed all samples for SAMe by the method of Baldessarini and Kopin (1) as modified by Matthysse and Baldessarini (6).

Figure 1 shows the blood concentrations of SAMe as a function of time and the two regression lines (α and β) from which we calculated the half-life value, $T\frac{1}{2}$. The double exponential pattern of the curve indicates that the substance is distributed according to a two-compartment model. The $T\frac{1}{2}$ value of 121 min (second phase, β) shows that the

---

[1] Neurospychiatric Hospitals, Verona (Italy).
[2] Department of Pharmacology, University of Milan (Italy).

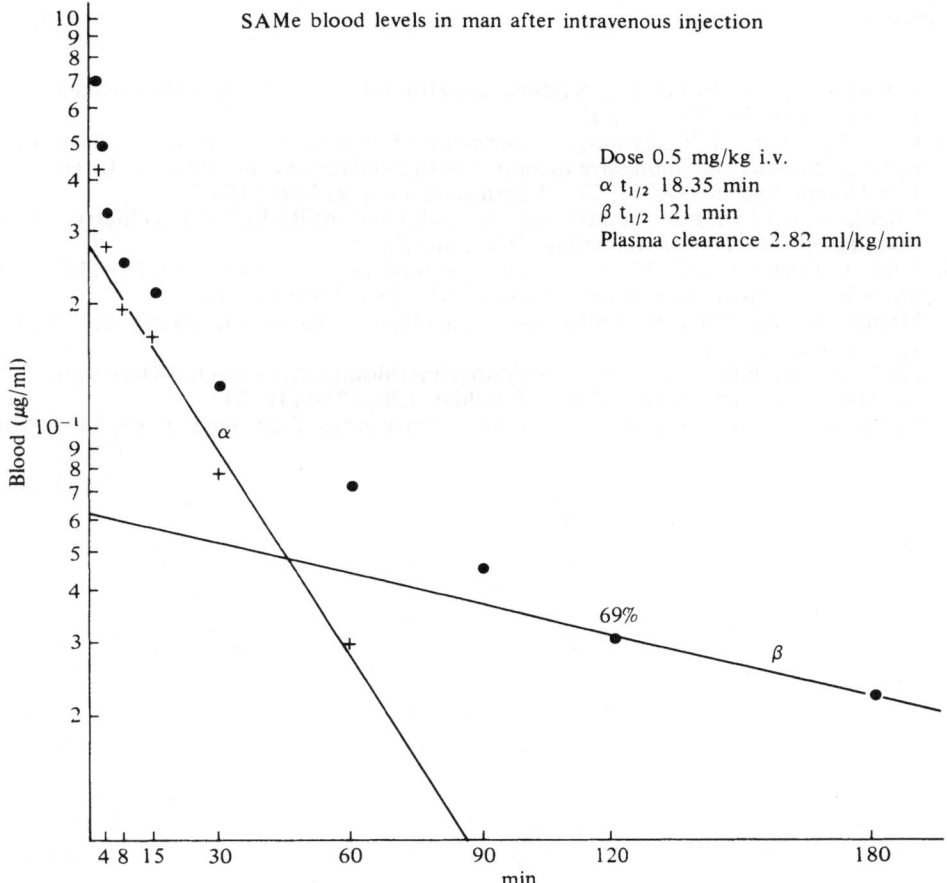

Fig. 1. SAMe blood levels as a function of time after intravenous dosing in man (0.5 mg/kg), and regression lines α and β

substance is metabolized rapidly, 69% of the administered dose being already metabolized and excreted 2 h after administration. This is confirmed by a high value of plasmatic clearance, namely 2.82 mg/kg/min, starting at equilibrium with the tissues and throughout elimination of the remaining 31% of the dose. The absorption $T\frac{1}{2}$ (first phase, α) is 18.35 min.

The remaining pharmacokinetic parameters of the two-compartment model (blood-tissues) are as follows:

Vc 155 ml/kg
Vd (area) 493 ml/kg
Kcl 0.01819 min^{-1}
K^{21} 0.01182 min^{-1}
K^{12} 0.01345 min^{-1}

146

References

1. Baldessarini, R.J., Kopin, I.J.: S-adenosylmethionine in brain and other tissues. J. Neurochem. **13**, 769 (1966)
2. Fazio, C., Andreoli, V., Agnoli, A., Casacchia, M., Cerbo, R., Pinzello, A.: Therapy of schizophrenia and depressive disorders with S-adenosyl-L-methionine. Intern. Res. Comm. System (IRCS), Clin. Pharmacol. Ther. **2**, 1015, 1974
3. Friedhoff, A.J.: Biogenic amines and schizophrenia. In: Biological Psychiatry. Mendels, J. (ed). New York: Wiley, 1973, pp. 113-129
4. Labò, G. Gasbarrini, G., Miglio, F.: Alcuni effetti delle transmetilazioni SAMe-dipendenti in epatologia. Minerva med. (Torino) **63**, 2007 (1972)
5. Mantegazza, P., Müller, E.: Relazione farmacologica sulla tossicità della S-adenosyl-metionina (in press)
6. Matthysse, S., Baldessarini, R.J.: S-adenosylmethionine and catechol-O-methyl-transferase in schizophrenia. Am. J. Psychiat. **128**, 1310 (1972)
7. Smythies, J., Antun, F.: The biochemistry of psychosis. Scot. med. J. **15**, 34 (1970)

S-Adenosyl-L-Methionine (SAMe) Blood Levels in Schizophrenia and Depression

V.M. ANDREOLI[1] , F. MAFFEI[1] , and G.C. TONON[2]

S-adenosyl-L-methionine (SAMe) serves as the methyl group donor for several transmethylase enzyme systems, the best known being catechol-O-methyltransferase (COMT), whose intracellular distribution is similar to that of SAMe (3); many other transmethylation reactions, however, are SAMe-dependent (6).

These reactions come into special prominence in the case of the CNS since many molecules occurring naturally in the brain, once methylated, acquire very peculiar properties in regard to behavior: thus N-methylated tryptamine derivatives, 5 OH- and N-methylated serotonin derivatives, and methylated dopamine all have definite psychotogenic effects (13, 14, 22). Also a variety of exogenous molecules, once methylated or methoxylated, may acquire psychotomimetic properties on behavior, a signal example being methoxylated amphetamine (1, 2). Then if we look at the biochemical interpretations of schizophrenia and consider the importance of the "transmethylation theory" (5, 13, 16), advocating the formation of endogenous methylated derivatives of catechols and indoleamines with psychotogenic effects, we can see the contributory value of studying transmethylation processes in which SAMe is involved as the donor of methyl radicals — the other two elements of the triad being the transmethylating enzyme and the substrate.

Previous studies of SAMe in the CNS indicate that this substance has a characteristic distribution both in a broad topographic sense (9) and within the neurons (3); further, there is good evidence that these cerebral concentrations of SAMe can be modified by the administration of methionine or methyl acceptors (7, 23) as well as by treatment with antidepressant drugs (8).

There are in addition clinical reports of aggravation of acute symptoms of psychosis in schizophrenic patients treated orally with methionine plus an MAO inhibitor (15, 21), or even with methionine alone (4). That this aggravation is caused by the activation of methionine to SAMe, and not to a nonspecific toxic action of methionine as such, is apparently demonstrated by the marked deterioration of symptoms observed in three schizophrenic patients after an intravenous injection of SAMe (12).

As for depression, here too an alteration of cerebral amine metabolism has been advocated as a pathogenetic factor (20); and a reduction of erythrocyte COMT activity was actually demonstrated (10). This last finding might reflect an absolute or relative dificiency of the methyl donor, as suggested also by clinical reports of significant improvement of depression in patients treated with SAMe (12).

[1]   Neuropsychiatric Hospitals, Verona (Italy).
[2]   Department of Pharmacology, University of Milan (Italy).

All these premises indicate that the biochemical study of SAMe may yield extremely interesting results in regard to the pathogenesis of certain behavioral disorders; and this is why we undertook an investigation of SAMe blood levels in normal subjects and in patients with depression or schizophrenia. In the last-named instance, Matthysse and Baldessarini (18) did a similar research in a group of chronic hospitalized schizophrenics but found no significant differences from normal values.

Material and Methods

We assayed SAMe blood levels in 15 normal subjects, 6 patients with depression, and 12 schizophrenics, by the method of Baldessarini and Kopin (9) as modified by Matthysse and Baldessarini (18). We tried to make all groups homogeneous and comparable by taking only male subjects and restricting the age spread to between 25 and 50 years. The patients were psychiatric hospital guests and during their hospital stay received all laboratory tests needed to detect liver, kidney, or cardiovascular disorders. We excluded from our study all subjects presenting abnormal blood chemistry findings. Of the 12 patients with schizophrenia, 5 had acute psychosis and 7 were chronic cases.

We collected all blood samples consistently at 10 a.m., as recommended by Matthysse et al. (19), in order to minimize possible scattering of SAMe assay returns due to circadian oscillations.

We extended our study of physiologic SAMe concentrations to include the urine and cerebrospinal fluid: this we did in a group of presumably normal subjects admitted to our neurologic wards for diagnostic procedures.

Results and Discussion

Table 1 shows the normal concentrations (μg/ml) of SAMe in the biological fluids of our control subjects. The total blood quota (0.88 ± 0.04 μg/ml) is divided between the plasma (0.46 ± 0.07 μg/ml, or 59%) and the blood cells (0.38 ± 0.05 μg/ml). The high urinary excretion (7.10 ± 0.35 μg/ml) found in our subjects was associated with normal water diuresis and creatinine clearance values.

Table 2 shows SAMe blood levels in the normal control group, in the depressed patient group, and in the schizophrenics. There was a statistically significant ($P < 0.001$) reduction of SAMe concentration in the blood of acute schizophrenics, but not in chronic schizophrenics or the depressed patient group.

Our results in normal controls and chronic schizophrenics agree with those of Matthysse and Baldessarini (8), who also failed to detect abnormal SAMe assays in the blood of chronic schizophrenics. The authors just named, however, did not investigate acute schizophrenics — in which group we did find a significant reduction of blood SAMe. This finding might be adduced in support of the hypothesis of an increase of transmethylation processes in the pathogenesis of schizophrenia: in other words, SAMe would be depleted through excessive consumption.

Before drawing any final conclusion, however, one should proceed from mere blood level studies to an investigation of SAMe turnover, and particularly of the balance between

Table 1. SAMe concentrations (μg/ml) in the blood, cerebrospinal fluid (CSF), and urine of normal subjects

Biological fluid	SAMe assay (μg/ml)	No. of subjects
Blood	0.88 ± 0.04	15
CSF	0.0128 ± 0.0007	5
Urine	7.10 ± 0.35	7

Values are given as means ± se.

Table 2. SAMe concentrations (μg/ml) in the blood of normal controls, depressed patients, and acute and chronic schizophrenics

			Schizophrenics	
	Controls	Depressed	Acute	Chronic
	0.88 ± 0.04	0.99 ± 0.17*	0.49 ± 0.03**	0.99 ± 0.04*
No of cases	15	6	5	7

* not significant
** $P < 0.001$
Values are given as means ± se.

SAMe and SAHC (S-adenosylhomocysteine), where SAHC inhibits the same transmethylation processes that are promoted by SAMe (11).

Quite recently, in fact, Levi and Waxmann (17) have demonstrated a reduced turnover of SAME in schizophrenic patients, and proposed an alternative theory advocating a blockade of transmethylation processes in schizophrenia.

References

1. Andreoli, V.: Anfetamine, loro derivatie comportamento. In: Quaderni die Neuropsicofarmacologia Andreoli, V., Del Mastro, S. (eds.) Societa italiana neuropsicofarmacologia Milano: 1971
2. Andreoli. V., Danieli, B., Tonon, G.C.: Significato dei derivati metossilati nelle psicosi da anfetamina. Riv. Farm. Ter. 4, 1a (1973)
3. Andreoli, V.M., Maffei, F., Tonon, G.C.: The subcellular distribution of S-Adenosyl-L-Methionine (SAMe) in the rat brain. This volume, p. 114
4. Antun, F.T., Burnett, G.B., Cooper, A.J., Daly, R.J., Smythies, J.R., Zealley, A.K.: The effects of L-methionine (without MAOI) in schizophrenics. J. Psychiat. Res. 8, 63, (1971).
5. Antun, F., Eccleston, D., Smythies, J.R.: Transmethylation processes in schizophrenia. In: Brain Chemistry and Mental Disease, New York: Plenum 1971 pp. 61-71
6. Axelrod, J.: Methyltransferase enzyme in the metabolism of physiologically active compounds and drugs. In: Handbook of Experimental Pharmacology. Eichler, O., Farah, H., Herken, H., Welch, A.D. (eds.) New York: Springer, 1971, Vol. XXVIII/2, p. 610

7. Baldessarini, R.J.: Factors influencing S-adenosyl-methionine levels in mammalian tissues. In: Amines and Schizophrenia. Himwich, H., Kety, S., Smythies, J. (eds.) Oxford: Pergamon Press, (1966), pp. 198-206
8. Baldessarini, R.J.: Alterations in tissue levels of S-adenosylmethionine. Biochem. Pharmacol. **15**, 741 (1966)
9. Baldessarini, R.J., Kopin, I.J.: S-adenosylmethionine in brain and other tissues. J. Neurochem. **13**, 769 (1966)
10. Cohn, C.K., Dunner, D., Axelrod, J.: Reduced catechol-O-methyltransferase activity in red cells of women with primary affective disorders. Science **170**, 1323 (1970)
11. Deguchi, T., Barchas, J.: Inhibition of transmethylations of biogenic amines by S-adenosylhomocysteine. J. biol. Chem. **246**, 3175 (1971)
12. Fazio, C., Andreoli, V., Agnoli, A., Casacchia, M., Cerbo, R., Pinzello, A.: Therapy of schizophrenia and depressive disorders with S-adenosyl-L-methionine. Intern. Res. Comm. System (IRCS), Clin. Pharmacol. Ther. **2**, 1015 (1974)
13. Friedhoff, A.J.: Biogenic amines and schizophrenia. In: Biological Psychiatry. Mendels, J. (ed.) New York: Wiley, 1973, pp. 113-129
14. Friedhoff, A.J., Schweitzer, J.W., Miller, J.: The enzymatic formation of 3, 4, di-O-methylated dopamine metabolites by mammalian tissues. Res. Comm. Path. Pharmacol. **3**, 293 (1972)
15. Kakimoto, Y., Sano, I., Kanazawa, A., Tsujio, T., Kaneko, Z.: Metabolic effects of methionine in schizophrenic patients pretreated with a monoamine oxidase inhibitor. Nature (Lond.) **216**, 1110 (1967)
16. Kety, S.S.: Current biochemical approaches to schizophrenia. New Engl. J. Med. **276**, 325 (1967)
17. Levi, R., Waxmann, S.: Epilepsy, cancer, and Schizophrenia. Lancet **II**, 11, 1975
18. Matthysse, S., Baldessarini, R.J.: S-adenosylmethionine and catechol-O-methyl-transferase in schizophrenia. Am. J. Psychiat. **128**, 1310 (1972)
19. Matthysse, S., Lipinski, J., Shih, V.: L-Dopa and S-adenosylmethionine. Clin. chim. Acta **35**, 253 (1971)
20. Mendels, J., Stinnett, J.L.: Biogenic amine metabolism, depression and mania. In: Biological Psychiatry. Mendels, J. (ed.) New York: Wiley, 1973, pp. 35-64
21. Pollin, W., Cardon, P.V., Kety, S.S.: Effects of amino acid feeding in schizophrenic patients treated with iproniazid. Science **133**, 104 (1961)
22. Saavedra, J.M., Axelrod, J.: Psychotomimetic N-methylated tryptamine formation in brain in vivo and in vitro. Science **175**, 1365 (1972)
23. Wurtman, R.J., Rose, C.M., Matthysse, S., Stephenson, J., Baldessarini, R.J.: L-Dihydroxyphenylalanine: effect on S-adenosylmethionine in brain. Science **169**, 395 (1970)

A Polygraph Evaluation of the Effects of S-Adenosyl-L-Methionine (SAMe) on All-Night Sleep

C. MAGGINI and M. GUAZZELLI

Classical screening procedures in animals seldom yield dependable indications about the therapeutic actions of a drug — particularly in the field of psychopharmacology; and yet these procedures are generally accepted for the characterization of new drugs.

On the other hand, as more and more new compounds are developed with potential therapeutic virtues, and animal experimentation contributes but little to their understanding, clinical pharmacology is increasingly called upon to provide vital information on such new drugs.

In the specific field of clinical psychopharmacology, these increasing demands have created the need for finer methods of evaluation and new technical shortcuts to limit the number of patients used in any given experiment, and also to save time and money (41). Conversely, such new methods must yield reliable information on drug activity in humans before the major committment of full-scale clinical trials with all their ethical and economic implications.

With these premises in mind, and considering the possibility of obtaining continuous information on cerebral function through monitoring of bioelectrical brain potentials, Itil and his colleagues (21) proposed the use of electroencephalographic techniques for the screening of new psychotropic drugs.

Qualitative and quantitative analysis of EEG records affords the recognition and definition of various types of bioelectrical reactions that are specific of each class of psychoactive and it affords definite models from established drugs to serve as terms of comparison.

Procedures of this type make it possible to detect resemblances between a new drug and one already in use, and conversely to define bioelectrical differences that indicate a new or peculiar therapeutic action. Sophisticated analytical techniques, and evaluation of the resting EEG records (quantitative pharmacoelectroencephalography), not only bring out the clinical effects of the test drug, thus predicting its potential therapeutic usefulness, but also afford identification of the effective dosage range (20).

These newly developed methods are associated to advantage with animal screening tests, providing as they do crucial knowledge to be added to the crop of classical experimental pharmacology; as a result, the new drug no longer reaches the stage of clinical experimentation with only a sketchy outline of its properties.

Like continuous EEG monitoring in the waking state, polygraphic recording of all-night sleep seems to answer certain requirements that are important not only in the screening stage but also in the clinical testing of a new drug (29).

Psychiatric Hospital, University of Pisa, Pisa (Italy).

In fact, in the course of treatments with psychoactive drugs, the all-night sleep record shows structural changes that make possible the differentiation of drugs belonging to different classes. Thus, for instance, benzodiazepines reduce stage 4 sleep and respect, or even slightly increase, REM sleep, whereas the barbiturates reduce both of these stages. Conversely, antidepressants increase stage 4 and reduce REM sleep (35).

Psychiatric patients invariably present altered sleep patterns; unfortunately, these alterations are not straightforward, possibly because of different criteria of selection, different dosages, and different routes of administration — but above all, we suspect, because of individual differences of basal psychopathology and sleep rhythms. For screening purposes, therefore, it would seem better to use normal subjects, whose sleep pattern modifications afford easy detection of the effects of psychotropic drugs. Also of practical importance is the fact that "it may be easier to find consenting volunteers than patients, when a laborious and perhaps distressing procedure is to be carried out" (28).

Also in the stage of clinical experimentation of a new drug it is most desirable to have, in addition to the experimenter' s own observations, the returns of a technique such as polygraphic recording, which provide objective evidence of the drug's biological activity, integrating subjective symptoms and broadening the scope of the investigation quite considerably. These new data promise a more accurate characterization of the test product and provide an answer to many questions such as the type of activity exerted, its degree of specificity, its intensity, and so on. Studies conducted with antidepressant and anxiolytic drugs in neurotic and endogenously depressed patients have revealed a close correlation between the improvement of depression and anxiety and some peculiar modifications of sleep, indicating that continuous polygraph recordings might constitute a serviceable instrument for the investigation of new drugs in psychiatric patients (3, 29). Similar studies carried out in schizophrenics treated with thioridazine have shown that phenothiazine therapy produces a significant improvement of schizophrenic symptoms associated with a parallel tendency to normalization of certain sleep disorders (42). Also, even though currently available knowledge does not pin down one or another alteration of the sleep record as specific to one or another psychopathologic entity, the constant finding of similar EEG alterations of sleep in various samples of patients selected according to identical criteria suggests the existence of a relationship between symptom features and polygraph findings. These clinical-EEG correlations again suggest that polygraphic records of the patient's all-night sleep might be taken as an index of symptom evolution; and that this characteristically objective method might do an excellent service in the clinical study of psychoactive drugs. This type of investigation seems to fulfill two essential requirements of clinical psychopharmacology: namely, to obtain a faithful record of the biological phenomenon, and at the same time to know that the record is closely connected with the clinical picture and its variations (29).

Thus, while "the doctor at the bedside remains the ultimate judge" (13), adoption of these methods will certainly enlarge the scope of psychoactive drug research; and it will also expedite and refine the characterization of new drugs awaiting clinical trials.

---

[1] SAMe, a natural donor of methyl radicals, is formed in vivo from the essential amino acid methionine, by the agency of the enzyme S-adenosyltransferase. The product, now made industrially in a stable form, was kindly supplied by the BioResearch Company of Milan.

In view of all these premises, therefore, we found it worthwhile to explore the effects of S-adenosyl-L-methionine (SAMe)[1] on the sleep pattern in terms of all-night polygraphic recordings, the better to characterize the drug and obtain inducations about its therapeutic potential — the substance being a likely putative psychoactive drug on account of its ability to modify the synthesis and catabolism of neuronal mediators in the CNS.

Material and Methods

This was a two-step investigation (Fig. 1). In the first stage we used three male volunteers (mean age 25.5 years), who were medical undergraduates. In the second stage, on the strength of the results obtained in the first run, we extended our exploration to two male patients hospitalized in our wards for endogenous depression: Romeo L., 67 years of age, suffering from inhibited unipolar depression and Renato F., 66 years of age, suffering from anxious unipolar depression. Both patients had a record of previous hospital stays in our clinic for depression as just specified, and both had proved uncommonly resistant to antidepressant medication. Practical difficulties prevented us from working with a larger sample and from finding younger patients for the trial.

The healthy volunteers were of similar cultural and socioeconomic background and were selected on the basis of a normal sleep-waking rhythm and after a physical checkup and a psychiatric interview. All three had taken no drugs for at least 1 month at the time of the trial; they were also instructed to abstain from all unusual activities and particularly not to take afternoon naps.

The trial consisted of 13 consecutive all-night polygraphic recordings. The first night was allowed for adaptation to the experimental situation and was disregarded; the second and third nights were used for recording each subject's basal sleep pattern. On the fourth day of the trial, SAMe was administered intravenously at a dosage of 30 mg immediately before the start of all-night recording. On the fifth night no treatment was given and the all-night record was studied for residual or withdrawal effects. Starting on the morning of the sixth day, each participant was given an intramuscular injection of 15 mg of SAMe as soon as they woke up; this was repeated for 6 consecutive days. All-night recordings were taken on these six nights and for two additional nights thereafter. Each morning, upon awakening, the subjects gave their own evaluation of the night's sleep by answering a specially prepared questionnaire.

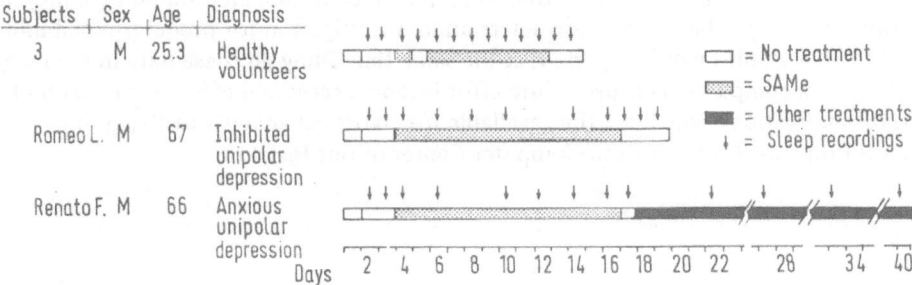

Fig. 1. Experimental design

Before the trial, each subject was given Cattel's Culture-Free General Intelligence Reactive test, Scale 3, and a California Psychological Inventory (CPI). Only the personality questionnaire was repeated at the end of the trial.

As for the two depressed patients participating in the trial, the criteria of selection were: absence of clinical evidence of CNS involutional pathology, and freedom from functional or organic pathology of the various organs and apparatuses.

After a 10-day period of pretreatment with a placebo, we made three all-night recordings, the last two of which were taken as basal values. Then we started intramuscular treatment with SAMe, the dosage being 45 mg daily in one patient (Romeo L.) and 45-90 mg daily in the other (Renato F.). All-night recordings were made as shown in Figure 1. The first patient (Romeo L.) was given no further treatment in connection with the trial. The second (Renato F.) was given 1 day's rest at the end of SAMe treatment and then was placed on antidepressant and anxiolytic therapy (sulpiride 300 mg, lorazepam 7.5 mg, imipramine 75 mg) plus a course of seven electroshock treatments. All-night polygraphic recordings were performed 1 day after the first, third, fifth, and seventh electroshock application (Fig. 1).

Clinical symptoms were assessed by administration of Hamilton's Rating Scale for Depression (RSD) on the same days as the all-night recordings were made.

The polygraphic study was carried out in the EEG Laboratory for the Study of Sleep in our own Department of Psychiatry. The patients were removed to the laboratory ½ h before their usual bedtime; recording was continued for 8 consecutive hours. We used five bipolar EEG leads plus two for recording eye movements and one for surface electromyography. Sleep stages were evaluated for periods of 20 s according to the classification of Dement and Kleitman (8). The following parameters were considered: Sleep Latency 1 (time lapsed between the start of recording and the onset of Stage 1); Sleep Latency 2 (time lapsed between the start of recording and the onset of Stage 2); Number of Awakenings their Duration; Final Awakening (duration of wakefulness to the end of recording time); Total Waking and Sleep Time; Sleep Stages (in absolute time values and as percentages of Total Sleep Time); Number of REM periods; Latency of First REM period (from sleep onset to First REM period).

Polygraphic data were processed statistically by variance analysis, discriminant analysis after Veldman (46), and stepwise discriminant analysis after Dixon (10); we utilized the automatic EEG data managing system in use in the Psychiatric Institute of the Univ. of Pisa. To that end, a form for recording the main EEG and clinical data from each patient and noting sleep stages over 20-s periods was prepared. Thus all the data gathered in an 8 h continuous recording were transferred to 25 punch cards and elaborated by a program that synthesized in a table all the parameters under investigation for predetermined intervals and for the whole recording period, at the same time filing all these data in the memory system of a computer. This procedure affords rapid access and effective retrieval of groups of observations, which are thus available for statistical analysis utilizing the Statistical Programs Library of the Computer Center of our Institute.

Results

Healthy Volunteers Study

1. *Polygraphic Investigation.* The basal sleep pattern was similar to those reported by Williams and his associates (50) for healthy volunteers of the same age group, particularly in regard to the incidence of Stage 4 and REM sleep.

With intravenous administration of 30 mg of SAMe an increase of Stage 4 and a reduction of REM sleep were observed. The remaining parameters were not remarkably modified. During the next night (fifth day of the experiment), Stage 4 showed a reduction and REM sleep an increase, with a tendency to restoration of pretreatment values.

The changes observed with intramuscular doses of SAMe (15 mg daily for 6 consecutive days) faithfully duplicated those seen with the single intravenous dose: Stage 4 was consistently increased and REM sleep was reduced. We also detected an increase of latency of the first REM period; other parameters were not remarkably affected, barring a slight increase of the number of awakenings and their duration during the third night of the treatment (eighth day of the experiment) (Table 1).

Figure 2 displays the effects of SAMe on Stage 4 and REM sleep: the former continuing above basal values throughout the trial, and the latter, conversely, continuing below pretreatment values.

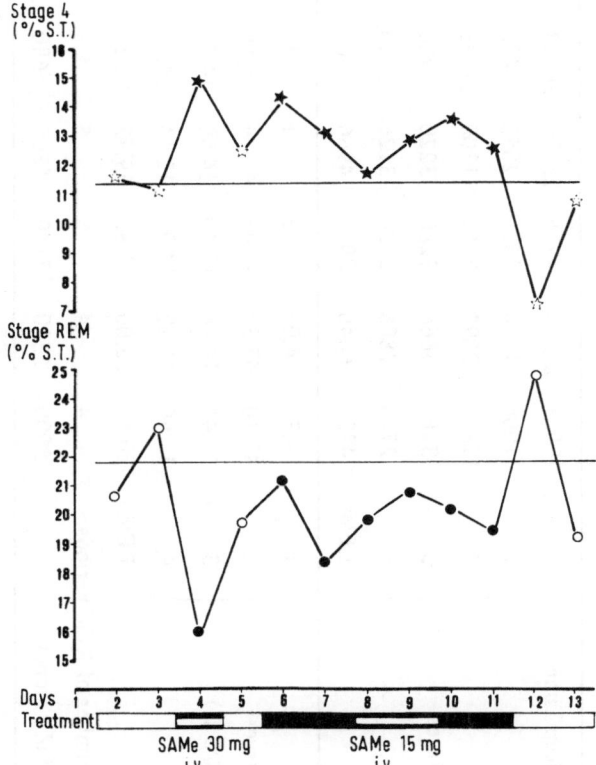

Fig. 2. Healthy volunteers study. Stages 4 and REM changes during the trial. Solid line indicetes the mean of baseline nights 2 and 3

156

Table 1. Healthy volunteers study. Mean values of sleep parameters at various observation times

Parameters*	Baseline		SAMe i.v.	SAMe withdrawal			SAMe i.m.			SAMe withdrawal		
Days	2	3	4	5	6	7	8	9	10	11	12	13
Sleep Latency 1	1126	2113	1333	820	666	700	1006	1360	1453	693	1666	1386
Sleep Latency 2	1420	2493	1973	1186	1000	1028	1280	1713	3000	1013	1920	1666
Awakenings (duration)	93	0	920	360	266	146	3000	500	120	0	0	213
Awakenings Number	0.6	0	1.6	2	0.3	1	3.3	1	0.6	0	0	0.6
Final Awakenin	4440	1166	1780	2533	1346	2326	1966	1580	1306	860	1353	2433
Total Wake Time	5460	3280	4000	3346	2286	3173	4973	3440	2880	1553	3060	4033
Sleep Time	23053	25513	24776	25086	26513	25626	23826	25360	25920	27246	25740	24766
Stages 1	1006	1046	1520	2053	1433	1453	1686	1993	1620	1300	1080	1220
Stages 2	19886	12093	12253	11300	11793	12306	10826	11853	13066	12820	12566	12873
Stages 3	3486	3666	3280	3220	3920	3846	3933	3040	2566	4520	3933	3460
Stages 4	2740	2806	3726	3126	3760	3333	2820	3253	3533	3453	1833	2493
Stages REM	4720	6640	3986	4966	5606	4686	4560	5220	5133	5153	6266	4720
Stages (% S.T.) 1	4.31	4.09	6.52	8.41	5.41	5.61	6.96	7.90	6.27	4.77	4.10	4.90
Stages (% S.T.) 2	47.10	47.32	49.43	44.74	44.56	48.39	44.92	46.82	50.34	49.96	48.75	52.16
Stages (% S.T.) 3	16.44	14.46	13.40	14.68	14.72	14.98	16.56	11.93	9.91	16.58	15.52	13.95
Stages (% S.T.) 4	11.60	11.12	14.98	12.44	14.17	12.99	11.74	12.82	13.63	12.66	7.16	10.11
Stages (% S.T.) REM	20.53	23.00	15.83	19.72	21.12	18.01	19.50	20.52	19.84	19.02	24.46	18.87
Number of REM Periods	4	4	5	5	5.3	5.6	6	4.3	4	4.3	5.3	5
REM Latency	5000	5360	7140	5360	6680	6106	4926	6466	8373	6440	5513	4326

* Times are in seconds.

In the first of the two all-night recordings made after discontinuation of SAMe treatment (12th day of the experiment) there was a reduction of Stage 4 to less than basal values, and increase of REM sleep to more than basal values, with only negligible alterations of sleep time and other parameters. The next night (13th day of the experiment) there was a sharp reduction of REM sleep, while Stage 4 almost reached pretreatment values.

Dixon's stepwise discriminant analysis (10) afforded differentiation of sleep patterns at various stages of the trial (Fig. 3).

The small size of our sample allowed only a comparison of grouped baseline and no-treatment data versus data obtained during SAMe treatment (Table 2).

Variance analysis of the data so grouped revealed that the increase of Stage 4 was statistically significant both in seconds and as a percent of ST ($P < 0.005$ and $P < 0.01$ respectively). Also significant was the increased latency of the first REM period ($P < 0.01$).

Veldman's discriminant analysis (46) produced significant values for the duration of awakenings (0.48), for Stage 4 in seconds (0.45) and as a percent of ST (0.49), and for latency of the first REM period (0.40).

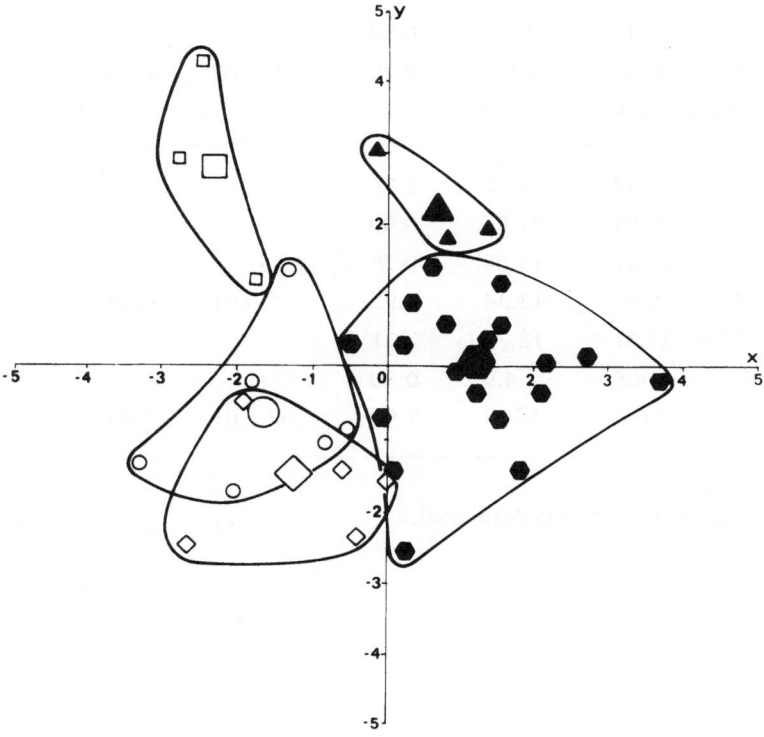

Fig. 3. Healthy volunteers study. Stepwise Discriminent Analysis results. Baseline: *white circles;* SAMe i.v.: *black triangles;* SAMe i.v. withdrawal: *white squares;* SAMe i.m. *black hexagons;* SAMe i.m. withdrawal: *white squares (oblique)*

Table 2. Healthy volunteers study. Mean values and variance and discriminant analysis results. (Baseline and No-treatment versus SAMe)

Parameters*		Mean values		F ratio	p	Discriminant function**
Sleep Latency 1		1416	1122	1.13	–	–
Sleep Latency 2		1737	1517	0.54	–	–
Awakenings (duration)		133	539	2.49	–	0.48
Number Awakenings		0.6	1.0	0.60	–	–
Final Awakening		2393	1623	3.21	–	–
Total Wake Time		3942	3285	1.28	–	–
Sleep Time		24857	25514	1.28	–	–
	1	1281	1548	1.95	–	–
Stages	2	11944	12208	0.16	–	–
	3	3716	3536	0.32	–	–
	4	2600	3411	9.29	< 0.005	0.45
	REM	5316	4810	1.22	–	–
	1	5.16	6.10	1.36	–	–
Stages	2	48.01	47.87	0.00	–	–
(% S.T.)	3	15.01	13.88	0.76	–	–
	4	10.49	13.34	7.93	< 0.01	0.49
	REM	21.31	18.79	2.21	–	–
Number of REM periods		4.6	4.9	0.30	–	–
REM Latency		5112	6706	7.99	< 0.01	0.40

*   Times are in seconds.
**   Only significant values ($\geqslant \cdot$ 40) are desplayed.

Table 3. Healthy volunteers study. Mean values and variance analysis results per hour of recording. (Baseline and No-treatment versus SAMe)

		First hour				Second hour		
		Mean values		F ratio	p	Mean values		F ratio	p
Sleep Time		2178	2442	0.88	–	3600	3592	1.45	–
	1	325	432	2.42	–	102	83	0.24	–
Stages	2	546	547	0.00	–	1340	1342	0.00	–
	3	468	423	0.26	–	654	535	0.71	–
	4	838	1039	0.82	–	1212	1471	0.94	–
	REM	–	–	–	–	290	195	1.96	–
	1	22.53	22.29	0.00	–	2.85	2.35	0.21	–
Stages	2	25.40	24.09	0.11	–	37.22	37.38	0.00	–
(% S.T.)	3	20.44	16.37	1.04	–	18.18	14.89	0.70	–
	4	31.61	37.23	0.61	–	33.66	40.93	0.96	–
	REM	–	–	–	–	8.07	4.43	1.94	–

		Third hour				Fourth hour		
Sleep Time		3600	3578	1.49	–	3578	3558	0.36	–
	1	101	119	0.23	–	137	158	0.21	–
Stages	2	1925	1708	1.56	–	2017	1988	0.01	–
	3	905	709	0.46	–	245	457	1.90	–
	4	149	442	5.60	< 0.05	265	195	0.25	–
	REM	518	509	0.00	–	913	759	0.67	–
	1	2.81	3.35	0.27	–	3.86	4.51	0.24	–
Stages	2	53.48	47.76	1.38	–	56.46	56.09	0.00	–
(% S.T.)	3	25.14	22.28	0.43	–	6.81	12.70	1.91	–
	4	4.14	12.40	5.65	< 0.05	7.37	5.43	0.25	–
	REM	14.40	14.20	0.00	–	25.48	21.34	0.66	–

		Fifth hour				Sixth hour		
Sleep Time		3546	3439	0.73	–	3580	3507	0.75	–
	1	140	171	0.40	–	170	210	1.20	–
Stages	2	1880	2008	0.20	–	2112	1784	2.10	–
	3	482	455	0.02	–	557	438	0.47	–
	4	104	170	0.37	–	30	53	0.27	–
	REM	940	632	1.89	–	709	1020	1.62	–
	1	3.96	5.76	0.82	–	4.77	6.18	1.58	–
Stages	2	53.03	58.17	0.43	–	58.97	50.50	1.91	–
(% S.T.)	3	13.70	12.87	0.02	–	15.59	12.37	0.44	–
	4	2.98	4.75	0.33	–	0.87	1.49	0.25	–
	REM	26.31	18.42	161	–	19.78	29.44	1.92	–

		Seventh hour				Eighth hour		
Sleep Time		3244	3472	1.46	–	1529	1923	1.56	–
	1	216	220	0.00	–	88	152	1.35	–
Stages	2	1213	1798	5.53	< 0.05	909	1030	0.25	–
	3	233	228	0.00	–	169	199	0.06	–
	4	–	38	0.99	–	–	–	–	–
	REM	1581	1188	1.91	–	362	540	0.86	–
	1	6.60	6.65	0.00	–	5.08	8.17	0.70	–
Stages	2	35.34	52.25	5.35	< 0.05	45.71	48.26	0.06	–
(% S.T.)	3	7.21	6.39	0.04	–	7.68	8.54	0.03	–
	4	–	1.07	1.01	–	–	–	–	–
	REM	50.83	33.71	4.47	< 0.05	14.85	30.47	2.73	–

Times are in seconds.

Variance analysis of sleep parameters evaluated over 1 h recording periods revealed, during SAMe treatment, a significant increase of Stage 4 in seconds and as a percent of ST (P < 0.05) during the third hour, and a likewise significant reduction of REM sleep in seconds and as a percent of ST (P < 0.05) during the seventh hour, when there was also a significant (P < 0.05) increase of Stage 2 (Table 3).

2. Side Effects and Subjective Evaluation of Sleep. The day after the first SAMe treatment (fifth day of the experiment), all three subjects reported a feeling of fairly severe somnolence — two of them took a 1 h nap in the afternoon, instructions to the contrary notwithstanding.

One subject complained of headaches at 12 and 13 days of the experiment, i.e., in the first 2 days following discontinuation of treatment. The same subject reported light and uneasy sleep the night after the intravenous dose of SAMe (fifth day) and in the first night after discontinuation of the intramuscular treatment (12th day). The other two subjects, and this same subject the other nights of the experiment, slept soundly and restfully.

3. Psychodiagnostic Investigation. The intelligence quotients of the three subjects were 148 (95th percentile for university undergraduates), 138 (83rd percentile), and 124 (59th percentile) respectively.

The California Personality Inventory (CPI) revealed a peculiar and constant rise of Si Scale ratings in retests; this scale expresses inner security and involves factors and motivations that facilitate success in circumstances where autonomy and independence represent positive behavior. Whereas all three personality profiles already showed valid functional capacities both socially and intellectually before the treatment, final retesting revealed definite consolidation of superior mental capacities: in one case the change was more than 1 sigma.

Depressed Patients Study

1. Polygraphic Investigation. Romeo L. showed a profoundly altered basal sleep pattern with a Sleep Time of only about 100 min due to a large Number of Awakenings and an early Final Awakening (Table 4)). His sleep was almost exclusively Stage 1 and Stage 2, with Stage 3 accounting for only 1.98% of ST and REM sleep for 11.84% of ST. Stage 4 was nonexistent.

SAMe treatment did not modify the Sleep Time and produced insignificant changes of the various stages; in particular, Stage 4 remained absent during the entire trial. The only exception was Stage REM, which dropped sharply during treatment and showed a conspicuous rebound after discontinuation of treatment (Fig. 4).

Renato F. showed a pretreatment sleep pattern much curtailed by an early Final Awakening and many Awakenings, the latter being responsible for an increase of Stage 1. Both Stages 4 and REM were severely reduced.

During SAMe treatment an increase of Sleep Time, through delayed Final Awakening, was observed while the number and duration of Awakenings were not remarkably modified. Sleep pattern structure was characterized by a reduction of Stages 1 and 3 and an increase of Stages 2 and REM. Stage 4 maintained almost pretreatment values throughout the study. There was also a steady reduction of latency of the first REM period (Table 4, Fig. 5).

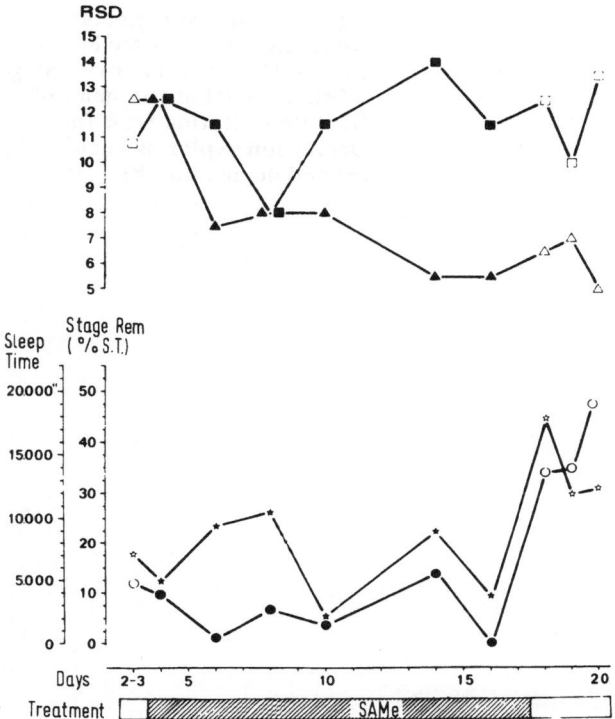

Fig. 4. Depressed patients study. Effects of SAMe in Romeo L.. Sleep time (★), stage REM (●), items of Hamilton's Rating Scale for Depression (RSD) exploring an anxiety (■) and depression (▲)

Compared to mean values elicited during treatment, discontinuation of SAMe medication produced a shortening of Sleep Time, an increased number of Awakenings and Stage 1, and a reduction of Stages 2 and 3. Stage 4 practically disappeared from the records, whereas REM sleep showed a marked increase not only relative to the treatment period but also to baseline values. The latency of the first REM period did not change from what it had been during treatment.

In this patient, further antidepressant and electroshock therapy resulted in some increase of Sleep Time, mostly reflecting a reduced duration of Awakening, and a reduction of Stage 1. Stages 4 and REM did not change remarkably from their respective values during SAMe treatment; also the latency of the first REM period remained the same. In the last two nights, coincident with a marked improvement symptomatology of depression, there was a sizable reduction of REM sleep and a slight increase of Stage 4 (Fig. 5).

2. Clinical Evaluation. So as to better explore the clinical effects of SAMe treatment, we grouped together the items of Hamilton's RSD exploring depression and those expressing anxiety, on the basis of a factorial analysis on all RSD items conducted in a large group of depressed patients (5).

Romeo L. showed a definite drop in score for the items exploring depression; score values for anxiety showed an initial fall and then a moderate increase. Discontinuation of SAMe

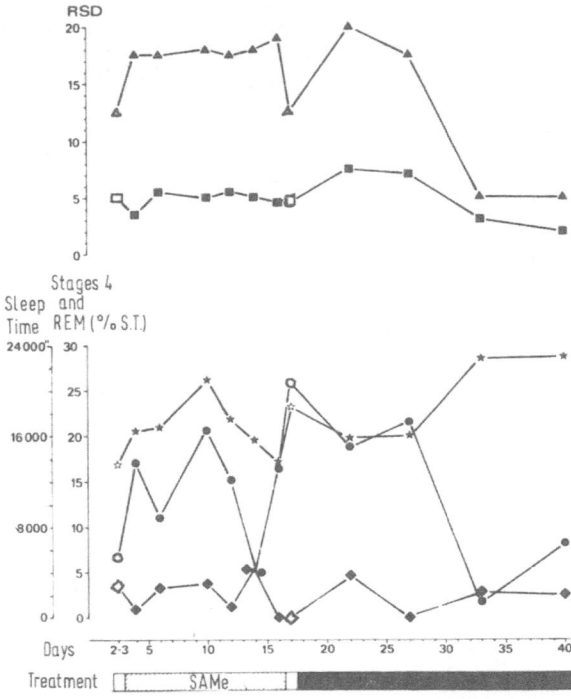

Fig. 5. Depressed patients
study. Effects of SAMe in
Renato F. . Sleep Time (★), Stage
4 (■), Stage REM (●), items of
Hamilton's Rating Scale for
Depression exploring anxiety
(▲) and depression (■)

treatment did not modify the overall profile of depression and anxiety ratings in any essential way.

Renato F. showed a rise of anxiety scores during SAMe medication, with essentially no changes in the remaining items; the anxiety scores dropped somewhat after discontinuation of treatment. Conventional therapy resulted in a substantial drop of RSD scores only after five sessions of electroshock therapy.

3. Clinical-Polygraphic Correlations. In Romeo L. we found a certain relationship between the reduction of RSD scores for depression and the reduction of Stage REM (Fig. 4). Conversely at 14 days of the experiment, coincident with a transient increase of stage REM above pretreatment values, there was an increase of score values for anxiety. Other than that, the mean value of Stage REM during treatment was only 5.88% of Sleep Time, i.e., appreciably less than pretreatment values; and this reduction of Stage REM was associated with a considerable increase of latency of the first REM period.

In the recordings made after discontinuation of SAMe treatment we detected a conspicuous rebound effect on REM sleep, which increased to better than 40% of Sleep Time and occurred very early in the night. These changes, however, occurred together with worsening of depression.

During SAMe treatment Renato F. failed to show noticeable changes of his clinical picture while his polygraphic recordings showed an increase of Stage REM and very low

Table 4. Depressed patients study. Mean values of some sleep parameters and of RSD items exploring anxiety and depression

Parameters[a]	Romeo L. Baseline	SAMe	SAMe withdrawal	Renato F Baseline	SAMe	SAMe withdrawal	Conventional treatment
Sleep Latency 1	5440	5480	1893	3286	–	1000	1935
Awakenings (duration)	10070	11905	12120	4970	5986	7080	2990
Number Awakenings	8.5	6.0	13.6	6.5	4.8	7.0	4.2
Final Awakening	6100	2210	773	7320	2740	2000	4575
Sleep Time	7190	6640	14013	13770	16786	18720	19300
Stages (% S.T.) 1	39.36	57.02	27.81	23.01	18.88	28.53	17.52
2	16.82	35.85	30.40	49.97	56.27	44.44	51.23
3	1.98	1.23	1.67	17.87	8.23	1.28	14.46
4	–	–	–	3.58	2.40	–	3.69
REM	11.84	5.88	40.11	6.85	14.18	25.75	13.08
Number of REM peroids	2.5	1.3	7.0	1.5	6.0	4	4
REM Latency	4110	9170	1813	9960	3966	3040	3450
RSD[b] Anxieta	10.7	11.5	12	12.5	17.9	12.5	11.8
Depression	12.5	7.8	6.1	5	4.8	4.5	4.8

[a] Times are in seconds.
[b] Anxiety includes the items: initial, central, and delayed insomnia, agitation, psychic, and somatic anxiety, gastrointestinal and general somatic symptoms. Depression includes: depression proper, guilt, suicide, work and hobbies, and retardation.

values for Stage 4. In the last two records made when the depression was practically under control, there was a definite drop of Stage REM both in seconds and as a percent of ST, while Stage 4 still remained very low (Fig. 5).

4. Side Effects and Subjective Evaluation of Sleep. Our two depressed patients showed no evidence of side effects attributable to SAMe treatment. Romeo L. said that his sleep was restful throughout the experiment, in contrast with objective findings indicating a severely impaired sleep. Renato F., instead, complained of unsatisfactory sleep all the time except in the last two nights of the treatment, coincident with a definite clinical improvement.

Discussion and Conclusions

Upon completion of our experimental work we are in a position not only to express a judgment on the effects of SAMe on the sleep pattern and offer a polygraphic profile that predicts a therapeutic action in clinical use, but also to make some firm points on the validity of sleep studies in psychoactive drug research. Our findings do indeed suggest the suitability of polygraphic sleep studies along with other methods for the screening of new psychoactive drugs, while at the same time demonstrating the usefulness of such studies in clinical experimentation, in addition to the experimenter's own evaluation.

Both as a single intravenous dose and in repeated intramuscular administration, SAMe showed the ability to produce important modifications of some parameters of sleep structure and organization in healthy volunteers. Quantitatively, however, sleep was essentially unchanged except for a slight and transient increase of Number of Awakenings. The parameters that showed statistically significant variations were Stage 4, Stage REM, and latency of the first REM period. Stage 4 showed an increase even with a single intravenous dose, and the increase was maintained throughout intramuscular treatment. The reduction of Stage REM, likewise observed after single and repeated dosage, was associated with an increased latency of the first REM period.

Thus SAMe appears to facilitate Stage 4 and inhibit REM sleep; all these effects abate shortly after discontinuation of treatment, with a rebound effect which is especially conspicuous in regard to REM sleep.

Overall, therefore, the polygraphic profile elicited in healthy volunteers brings out a resemblance between SAMe and antidepressant drugs, which produce very much the same modifications of sleep patterns. Many published reports indicate that tricyclic antidepressants and MAO inhibitors produce an increase of Stage 4 sleep (6, 7, 29, 35, 51) and a reduction of Stage REM (2, 7, 9, 11, 16, 17, 26, 29, 31, 35, 39, 40, 45, 48, 49,); there is also an increase of latency of the first REM period (16, 17, 35, 37), itself an expression of inhibition of paradoxical sleep.

The reduction of REM sleep differs according to the type of antidepressant drug used: with the tricyclic compounds it occurs promptly (with the very first dose), but it is usually transient and tends to abate with continuing treatment (25, 44); with MAO inhibitors, instead, the decrement is gradual, may go all the way to complete suppression of REM sleep, and will last for as long as the treatment is continued (37).

Discontinuation of antidepressant therapy invariably results in an increase of REM sleep, either simply to pretreatment levels (17) or beyond-implementing in that case a rugular "rebound" effect (12, 27, 37, 48).

In terms of its effects on REM sleep, SAMe seems to behave more like the tricyclic antidepressants than like MAO inhibitors the reduction of REM sleep is immediate and conspicuous (even with a single dose), and it never reaches complete suppression.

As we said before, the polygraphic profiles obtained in our healthy volunteers encouraged us to extend our research to two patients with active endogenous depression. Romeo L., suffering from an inhibited unipolar depression, showed a fair measure of clinical improvement with lower scores for items exploring depression and psychomotor inhibition, counterbalanced by a moderate increase of anxiety scores. Renato F., suffering from a unipolar anxious depression, showed no remarkable modification of items exploring depression, but rather an increase of the scores for agitation, psychic anxiety, general somatic symptoms, and insomnia. In this patient, however, a later therapeutic attempt with conventional antidepressant and anxiolytic drugs, continued for a fortnight, also resulted in mild clinical deterioration, and a reduction of RSD scores for anxiety and depression was achieved only after five sessions of electroshock therapy.

The limited results obtained in these two cases should not be construed as reflecting poor activity of the study product: both subjects were uncommonly resistant to psychotropic drugs, and in earlier hospital stays nothing short of ECT had been enough to abate their depression. The appreciable results obtained in Romeo L., whose more prominent symptoms were severe affective depression and psychomotor inhibition, suggest that SAMe might be particularly effective on those items.

Aside from clinical results, our polygraphic study of two depressed patients afforded only partial confirmation of the sleep pattern changes formerly observed in healthy volunteers. In particular, Stage 4 was totally suppressed in Romeo L. throughout the experiment, and remained essentially unchanged in Renato F. even after the resolution of clinical signs of depression.

Now, while it is generally recognized that severe reduction or even the complete suppression of Stage 4 is a prominent and typical alteration of the sleep pattern of depressed patients, no satisfactory interpretation of this phenomenon has been forthcoming. Some authors point out that Stage 4 tends to increase with the abatement of depression, and infer that the damping of Stage 4 is just another expression of the mechanisms that produce dysthymia (33, 34, 36). One step further, other investigators reason that since Stage 4 tends to remain below normal values compared to healthy agemates even after the abatement of depression, this paucity of Stage 4 sleep should be attributed not so much to the extant depression as to a more general biological setup characteristic of candidates to depression (18, 30) — or at least to the persistence of a physiopathologic arrangement of the sleep pattern which is inherent in dysthymia (32). At any rate, a definite connection between a dearth of Stage 4 sleep and depression is demonstrated not only by the constant reduction of this stage in spontaneously depressed patients, but also by the depression and hypochondria that develop in subjects artificially deprived of Stage 4 sleep (1). On the other hand, this relationship should be viewed with some circumspection, since impairment of delta sleep is by no means an exclusive appanage of dysthymia but occurs in other mental diseases, notably schizophrenia (4, 31).

Thus, if we bear in mind the erratic behavior of Stage 4 in depressive syndromes, we realize that the lack of effects from SAMe on this stage of sleep is not necessarily at variance with the data obtained in healthy volunteers. In addition to all that, we must consider the age of our two patients — and the physiologic shrinking of delta sleep that goes with increasing age. In a group of elderly subjects (between 71 and 95 years old), Kahn and Fisher (24) found an incidence of Stage 4 of as little as 4.5% of Sleep Time, with some subjects showing no trace at all of Stage 4.

Coming now to REM sleep, Romeo L. showed an inhibition of this stage while receiving SAMe treatment — more precisely, a moderate reduction of absolute and percentage values, and an increased latency of the first REM period each night; this was associated with clinical alleviation of depression. Renato F., who showed practically no change of depressive symptoms with SAMe, also failed to show any modification of REM sleep, either absolute or relative to Sleep Time; in this patient, indeed, a reduction of REM sleep occurred only after an additional fortnight of conventional antidepressant, anxiolytic, and ECT treatment. From these results it seems evident that a reduction of REM sleep occurs concomitant with the alleviation of depression, as pointed out by LeGassicke et al., (26) and Akindele et al. (2).

Hartmann (16) put forward an interesting idea to account for the apparently paradoxical behavior of REM sleep, which tends to decrease while all other stages tend to normalization. He says that the compensatory quantitative increase and reduced latency of REM sleep, which is constantly observed in subjects selectively deprived of it for some time, indicates a peremptory need for this stage of sleep, which he calls "need for D" or "pressure towards D." In depression, where the inherent pathophysiologic mechanisms of the CNS disturbance increase the number of Awakenings and their duration and reduce Sleep Time, there would be a true deprivation of REM sleep without any possibility of recoupment; and the pathologically increased "need for D" would contribute to the maintenance of depression by a feedback mechanism. According to Hartmann (16), the reduction of REM time induced by a drug could represent either a reduced need for that stage of sleep or a plain deprivation effect. Which of the two mechanisms is in play in any given case would be indicated by the behavior of REM sleep after discontinuation of treatment: a compensatory increase would indicate deprivation, whereas no increase would indicate a reduced need. In the author's interpretation, antidepressant drugs would act precisely by the latter mechanism, i.e., by breaking up the vicious circle in which "need for D" would stand as a perpetuator of depression.

Some later studies, however, failed to confirm this theory: both Wyatt et al. (48) and Passouant et al. (37) reported frequent compensatory increases of REM sleep after withdrawal of antidepressant medication.

While Hartmann's ideas (16) may be questioned, there is no denying that there is a close connection between REM sleep behavior and the alleviation of clinical depression, as is shown by Vogel's experiments of selective nonpharmacologic REM deprivation in depressed patients (47).

The results of our own work seem to show quite clearly that the polygraphic response to SAMe treatment depends on the quality of the basal sleep pattern. More precisely, whereas in normal subjects SAMe produced effects on sleep faithfully duplicating those of antidepressant agents, comparable changes were elicited in depressed patients only to a limited extent, and then only when the clinical picture of depression began to show some substantial changes. Accordingly, we must conclude that polygraph monitoring of psychiatric patients does not yield definite answers in the screening of a new psychoactive drug — especially with short treatments and limited periods of observation. In contrast, the data obtained from normal subjects appear far more reliable and may indeed constitute useful information toward the characterization of a new test drug.

In the stage of clinical experimentation, however, sleep studies certainly represent a useful adjunct to the experimenter's observations, affording as they do a better definition of drug properties and providing more complete information on the type, intensity, and specificity of the observed actions. Also, in the light of the biochemical theory of sleep stages (22, 23), it appears most desirable to integrate any investigation of clinical psychopharmacology with appropriate polygraphic studies, since all psychoactive drugs act upon the biochemical systems which underlie the bioelectrical mechanisms of sleep and at the same time, most probably, constitute the foundation of psychopathologic disturbances (38, 43).

To sum up, our polygraphic investigation of human sleep, applied for screening purposes to the study of SAMe, revealed changes of the sleep pattern quite comparable to those induced by antidepressant drugs. These results suggest the potential usefulness of SAMe as a therapeutic agent for depression, and at the same time indicate that the observed effects may reflect SAMe interference with the synthesis and catabolic disposal of CNS mediators (14).

References

1. Agnew, H.W., Jr., Webb, W.B., Williams, R.L.: Comparison of stage four and 1-REM sleep deprivation. Percept. Motor Skills **24**, 851, (1967)
2. Akindele, M.O., Evans, J.I. Oswald, I.: Mono-amine oxidase inhibitors, sleep and mood. Electroenceph. clin. Neurophysiol. **29**, 27 (1970)
3. Bedarida, D., Maggini, C., Riccioni, R., Castrogiovanni, P., Cassano, G.B.: Valutazione dell'azione ansiolitica ed euipnica di un nuovo derivato benzodiazepinico (ER 115). Indagine clinico-poligrafica. Atti 2a Riunione Naz. Soc. Ital. di Neuropsicofarmacologia, Tirrenia (Pisa), 14-16 June, 1969
4. Caldwell, D.F., Domino, E.F.: Electroencephalographic and eye movement patterns during sleep in chronic schizophrenic patients. Electroenceph. clin. Neurophysiol. **22**, 414 (1967)
5. Cassano, G.B., Castrogiovanni, P., Conti, L., Lucchelli, P.E., Sarteschi, P.: The computer data bank as a tool for the characterization of new antidepressants. IX C.I.N.P., Paris, 7-12 July 1974
6. Cazzullo, C.L., Penati, G., Bozzi, A., Mangoni, A.: Sleep patterns in depressed patients treated with a MAO inhibitor. Correlation between EEG and metabolites of Tryptophan. Com. al 6° Congr. Intern. C.I.N.P., Tarragona, 24-27 April, 1968
7. Cramer, H., Kuhlo, W.: Effects of inhibitors of MAO on sleep and EEG in man. Acta neurol. belg. **67**, 658 (1967)
8. Dement, W.C., Kleitman, N.: Cyclic variations in EEG during sleep and their relation to eyes movements, body motility and dreaming. Electroenceph. clin. Neurophysiol. **9**, 673 (1957)
9. Desmond, L.F., Dunleavy, M.D., Oswald, I.: Phenelzine, mood response and sleep. Arch. gen. Psychiat. **28**, 353 (1973)
10. Dixon, W.J.: Biomedical computer programs. Los Angeles: University of California: Berkeley. Press, 1968

11. Dunleavy, D.L.F., Brezinov, A.Y., Oswald, I., MacLean, A.W., Tinker, M.: Changes during weeks in effects of trycylic drugs on the human sleeping brain. Brit. J. Psychiat., 120: 663-672 (1972)

12. Dunleavy, D.L.F., Brezinova, V., Oswald, I., McLean, A.W., Tiker, M.: Changes during weeks in effects of tricyclic drugs on the human sleeping brain. Brit. J. Psychiat., 1972

13. Dunlop, P.: Principi informatori in una prova clinica di nuovi farmaci. Recent Progr. Med. **34**, 4, 1963

14. Fazio, C., Andreoli, V., Agnoli, A., Casacchia, M., Cerbo, R.: Effetti terapeutici e meccanismo d'azione della S-adenosil-L-methionina (SAMe) nelle sindromi depressive. Minerva med. **64**, 1515 (1973)

15. Deleted in proof

16. Hartmann E.: On the pharmacology of dreaming sleep (in D state). J. nerv. ment. Dis. **146**, 165 (1968)

17. Hartmann E.: Amitryptiline and imipramine: effects on human sleep. Report to the Association for the Psychophysiological Study of Sleep, 1968. Abstracted, Psychophysiology **5**, 207 (1968)

18. Hawkins, D.R.: Sleep, dreaming and clinical psychiatry. In: Sleep and Dreaming. Hartmann, E., (ed.) Boston: Little, Brown and Co., pp. 85-92, 1970

19. Deleted in proof

20. Itil, T.M.: Quantitative pharmaco-electroencephalography. (The use of computerized cerebral biopotentials in drug research). In: Psychotropic Drugs and the Human EEG (Itil, T.M. (ed.) Basel – New York: Karger, 1973, Vol. VII: Modern Problems of Pharmaco-Psychiatry

21. Itil, T.M., Guven, F., Cora, R., Hsu, W., Polvan, N.W., Ucok, A., Sanseigne, A., Ulett, G.A.: Quantitative pharmaco-electroencephalography using frequency analyzer and digital computer methods in early drug evaluations. In: Drug, Development and Brain Function (Smith, W.L., (ed.) Springfield (Illinois) Charles C. Thomas, 1971

22. Jouvet, M.: Mechanism of the states of sleep: A neuropharmacological approach. In: Sleep and Altered States of Consciousness (Kety, S., Evarts, E., Williams, H. (eds.) Baltimore: Williams & Wilkins 1967

23. Jouvet, M.: Biogenic amines and the state of sleep. Science **163**, 32 (1969)

24. Kahn, E., Fisher, C.: The sleep characteristics of the normal aged male. J. nerv. ment. Dis. **148**, 478 (1969)

25. Kramer, M., Whithman, R.M., Baldridge, B., Ornstein, P.H.: Drugs and dream. III) The effects of imipramine on the dreams of depressed patients. Amer. J. Psychiat. **124**, 1853 (1968)

26. LeGassicke, J., Ashcroft, G.W., Eccleston, D., Evans, J.I., Oswald, I., Ritson, E.B.: The clinical state, sleep and amine metabolism of a tranylcypromine (parnate) addict. Brit. J. Psychiat. **111**, 357 (1965)

27. Lewis, S.A., Oswald, I.: Overdose of tricyclic antidepressants and deduction concerning their cerebral action. Brit. J. Psychiat. **115**, 140 (1970)

28. Lucchelli, P.E., Del Mastro, S.: Procedura per lo studio clinico pianificato di un farmaco attivo sull'insonnia. Quaderni di Neuropsicofarmacologia, Milano (Italy), 29-30 June, 1 July, (1972)

29. Maggini, C., Gallevi, M., Conti, L., Placidi, G.F., Castrogiovanni, P.: Impiego dell' indagine poligrafica del sonno nello studio delle caratteristiche terapeutiche di un nuovo psicofarmaco (AF 1161). Boll. Soc. Med. Chir. Pisa, **4**, 294 (1969)

30. Maggini, C., Guazzelli, M., Conti, L., Mauri, M.: Polygraphic sleep study in neurotic and endogenous depressive patients. A multivariate statistical analysis approach. International Symposium on EEG and Sleep, Madrid, 9-11 May, 1974

31. Maggini, C., Murri, L., Pappagallo, S.: Il sonno notturno nelle sindromi schizofreniche. Lav. Neuropsichiat. **44**, 703 (1969)

32. Mendels, J., Chernik, D.A.: Psychophysiological studies of sleep in depressed patientis an overview. Presented at the Annual Conference on Current Concerns in Clinical Psychology, University of Iowa, Nov. 7-8, 1972

33. Mendels, J., Hawkins, D.R.: Sleep and depression, a follow-up study. Arch. gen. Psychiat. **16**, 546 (1967)
34. Mendels, J., Hawkins, D.R.: Sleep and depression (further considerations). Arch. gen. Psychiat. **19**, 445 (1968)
35. Muratorio, A., Maggini, C.: Psicofarmaci e sonno. Neuropsichiatria **35**, 1 (1969)
36. Muratorio, A., Maggini, C., Marcacci, G.: Evoluzione del sonno nelle sindromi depressive in corso di trattamento. Rass. Studi psichiat. **57**, 351 (1968)
37. Passouant, P., Cadilhac, J., Ribstein, M.: Les privations de sommeil avec mouvements oculaires par les anti-depresseurs. Rev. neurol. **127**, 173 (1972)
38. Kety, S.S.: Current biochemical approaches to schizophrenia. The new Engl. J. of Med. **276**, 6: 325-331 (1967)
39. Ritvo, E.R., Ornitz, E.M., La Franchi, S., Walter, R.D.: Effects of imipramine on the sleep dream cycle: an EEG study in boys. Electroenceph. clin. Neurophysiol. **22**, 465 (1967)
40. Ryba, P., Engelhardt, D., Freedman, N.: The effects of imipramine on sleep patterns of psychiatry patients. Report to the Association for the Psychophysiological Study of Sleep, Gainesville, March 1966
41. Sarteschi, P., Cassano G.B., Castrogiovanni, P., Conti, L.: L'iter di sperimentazione clinica di uno psicofarmaco: obblighi metodologici e soluzioni alternative. Atti IV Riun. Soc. Ital. di Neuropsicofarmacologia, Bologna (Italy), 23-24 Oct., 1971
42. Sarteschi, P., Cassano, G.B., Maggini, C.: Polygraphic study of all-night sleep in schizophrenics treated with thioridazine. Pharmacological and Clinical Aspects of Psychotropic drug Therapy, Moscow, 11-12 Sept., 1969
43. Schildkraut, J.J., Kety, S.S.: Biogenic amines and emotion. Science **156**, 21 (1967)
44. Tolle, R., Crome, A.: Zur Frage der Traumaktivierung durch Thymoleptica. Arzneimittel Forsch. (Drug Res.) **20**, 886 (1970)
45. Toyoda, J.: The effect of chlorpromazine and imipramine on the human nocturnal sleep electroencephalogram. Folia psychiat. neurol. jap. **18**, 198 (1964)
46. Veldman, D.J.: Fortran Programming for the Behavioral Sciences. New York: Holt, Rinehart and Winston, 1967
47. Vogel, G.W., Traub, A.C., Ben-Horin, P., Meyers, G.M.: REM deprivation. II) Effects on depressed patients. Arch. gen. Psychiat. **18**, 301 (1968)
48. Wyatt, R.J., Fram, D.H., Kupfer, D.J., Snyder, F.: Total prolonged drug-induced REM sleep suppression in anxious depressed patients. Arch. gen. Psychiat. **24**, 145 (1971)
49. Wyatt, R.J., Kupfer, D.J., Scott, J., Robinson, D.S.K., Snyder, F.: Longitudinal studies of the effect of monoamine oxidase inhibitors on sleep in man. Psychopharmacology **15**, 236 (1969)
50. Williams, R.L., Agnew, H.W., Webb, W.B.: Sleep patterns in young adults: an EEG study. Electroenceph. clin. Neurophysiol. **17**, 376 (1964)
51. Zung, W.W.K.: Effect of antidepressant drugs on sleeping and dreaming. Part I: On the adult female. Psychiatrists Association Meeting Atlantic City, N.J., May 1966

Results and Possible Developments of the Clinical Use of S-Adenosyl-L-Methionine (SAMe) in Psychiatry

A. AGNOLI[1], V.M. ANDREOLI[2], M. CASACCHIA[3], F. MAFFEI[2], and C. FAZIO[3]

SAMe Tolerability in Animals and Man

Experimental work on the acute, chronic, and fetal toxicity of SAMe was carried out by Stramentinoli and his colleagues (19). The intravenous LD_{50} in rats is more than 2000 mg/kg, or the equivalent of 140-150 g in a man of 70 kg. In man, on the other hand, clinical trials have revealed a definite action of the substance in the CNS with doses as little as 80 μg/kg; the largest dose so far administered intravenously to humans is 0.5 g daily and has produced no visible side effects.

Chronic toxicity tests, carried out in male and female rats, have shown no evidence of toxic effects: animal weight curves, organ weight, hematopoiesis, blood chemistry tests, and histologic examinations of various organs and tissues never differed significantly from the corresponding values obtained in untreated control animals.

Fetal toxicity tests in the rat and rabbit, with daily administration of SAMe at a dosage of 10 mg/kg throughout pregnancy, revealed no toxic effects on embryogenesis, fetal development, or fetal viability and morphology (19).

SAMe is formed naturally in the living body from methionine, itself a substance of clinical interest, at least one which was so for some time in the past.

Methionine is involved in the synthesis of protein and functions as a donor of methyl radicals − only however, after being converted to SAMe; because of this property, methionine was regarded in the past as a "lipotropic factor" − meaning one with the ability to reduce the accumulation of fat in the liver (21).

But the ingestion of methionine in amounts exceeding normal requirements by as little as three times causes severe toxic manifestations. Rats, guinea pigs, and rabbits treated with such dosages developed fatty degeneration of the liver to the point of liver necrosis, hypoglycemia to the point of coma, degeneration of the pancreas with loss of basophil staining affinity of the acini, and hypothermia eventuating in death (6, 7, 12, 13).

In man, severe methionine toxication "is virtually identical to that observed in animals." Man is in fact a species highly susceptible to the toxic effects of methionine, "developing acute toxic symptoms with dosages lower than those used in the guinea pig and lower still than those used in the rat" (11).

---
[1] Department of Nervous and Mental Diseases, University of L'Aquila (Prof. A. Agnoli).
[2] Neuropsychiatric Hospitals, Verona.
[3] First Department of Nervous and Mental Diseases, University of Rome (Prof. C. Fazio).

The mechanism of methionine toxication is not yet completely understood (8). The more generally accepted hypothesis stems from the discovery that methionine toxicity is accompanied by a commensurate depletion of ATP and gradual increase of SAMe and S-adenosylhomocysteine (SAHC): thus, methionine toxicity is apparently mediated by ATP deprivation, as is the case with ethionine toxicity (11).

It follows that the action of methionine depends on its previous conversion to SAMe; but this conversion requires such a consumption of ATP as to deplete liver reserves of the substance and upset the lipid, protein, and carbohydrate metabolism.

Since, however, SAMe levels in the blood serum are to a certain extent constant, and since SAHC is formed along with SAMe, it follows that the formation of SAMe from exogenous methionine again involves an inordinate expenditure of ATP. Thus the "active lipotropic factor" that subserves transmethylation processes in the liver is actually SAMe and not methionine (14, 20).

SAMe and Schizophrenia

In their original clinical work, Pollin and his colleagues (18) have demonstrated that the treatment of acute shizophrenics with methionine associated with MAO inhibitors produces an aggravation of symptoms. The same thing was confirmed in patients receiving methionine alone (3).

From these results it was inferred that methionine acted only after being converted to SAMe; and it was in fact demonstrated that treatment with large doses of methionine increased the concentration of SAMe in the brain (5).

In turn, the greater availability of methyl groups from brain SAMe would encourage the formation of methylated or methoxylated amines, in keeping with the transmethylation hypothesis advanced for schizophrenia.

We have ourselves treated three schizophrenic patients in the acute phase with SAMe (15 mg daily by intramuscular injection), and noted a deterioration of acute symptoms not only as a clinical impression but also as assessed by Wittenborn's symptom scale (10). The graphs (Fig. 1) show a constant reduction of weighted values for the fourth group of symptoms, concerning depression.

Next, we decided to look more analytically at the several items of depression before and after treatment by applying the McNemar test (since a normal distribution of separate score values cannot be assumed). And while a certain reduction of score values was apparent for all components, this reached statistical significance for item 13 (retardation and psychomotor activity).

Other than that, all three patients showed a particularly marked deterioration in terms of the items: anxiety episodes, paranoid schizophrenia, and manic symptoms. These results of SAMe treatment were closely comparable to those obtained by other investigators with methionine and MAO inhibitors.

From this preliminary evidence it seems permissible to conclude that the clinical effects observed with methionine treatment actually involved the conversion of methionine to SAMe.

On the other hand, SAMe might also act by an amphetaminelike mechanism of stimulation, which might have compounded the endogenous symptoms.

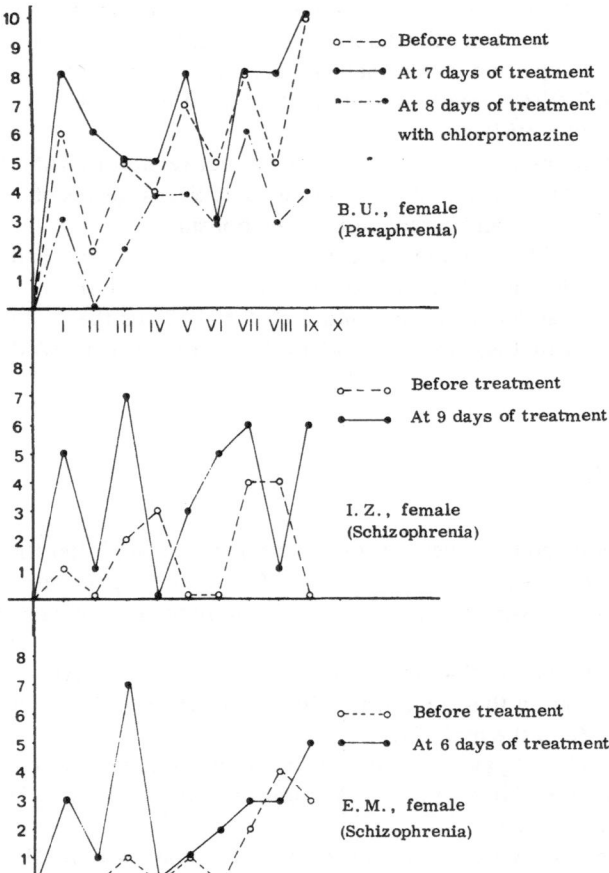

Fig. 1. Action of S-adenosyl-L-methionine (SAMe)
as assessed by Wittenborn's
scale in patients with acute
schizophrenia.
(From Fazio et al., 1974)
(10)

SAMe and Normal Subjects

In order to explore the possible stimulant effects of SAMe we made a special study of
10 normal volunteers. Before treatment, each participant was given a Minnesota Multi-
phasic Personality Inventory (MMPI) to define personality profiles and rule out pathologic elements. Subjects with extant cardiovascular or liver pathology were not included
in this study.

After this screening for psychic "normality," we gave each participant a self-rating
scale form to fill: namely the Mood Adjective Check List of Nowlys (17), which in
its reduced form is very easy to administer and evaluate, while covering a broad range
of moods. The check list consists of 33 adjectives, which the subject is asked to rate
with a score value from 1 to 5; the 33 adjectives are then summarized in 11 items obtained by a process of factorial analysis.

The test-retest reliability for the several items is variable but generally acceptable (between 0.40 and 0.70) for practical purposes. Our volunteers were given the test daily for 10 consecutive days before and after administration of 100 mg of SAMe daily by intravenous drip infusion. Table 1 shows the results. As can be seen, none of the items was influenced significantly by the treatment except asthenia ($P < 0.01$). We must point out, however, that the asthenia item responds significantly also to placebo (15).

From this evidence, therefore, it seems legitimate to conclude that SAMe is devoid of direct psychostimulant effects in normal subjects, at least in the conditions of this experiment.

SAMe and Depressive Syndromes

In view of the constant reduction of scores in the fourth group of symptoms in the Wittenborn scale (depression) in schizophrenic patients, and also of the modification of the "asthenia" item in normal subjects (this being a prominent symptom in depression), we decided to explore the effects of SAMe in depressive syndromes.

This avenue of clinical investigation was at the same time suggested by several hints in the literature. Thus Axelrod and his associates (4) detected a reduction of COMT activity in the red blood cells of depressed patients; and Baldessarini (5) showed that animals treated with tricyclic or MAO-inhibiting drugs (all of them antidepressants)

Table 1. Comparison of items in the Mood Adjective Check List of Nowlys in normal, nondepressed subjects, before and after a 10 day treatment with SAMe by intravenous drip infusion at a dosage of 100 mg daily

	t	Mean before treatment	Mean after treatment
Excitation	NS	1.60	1.15
Gaiety	NS	2.25	1.74
Egotism	NS	1.09	9.75
General activation	NS	3.45	2.76
Concentration	NS	3.97	3.57
Asthenia	0.01	2.42	1.43
Social participation	NS	4.79	4.53
Skepticism	NS	2.15	1.63
Sadness	NS	2.88	2.18
Anxiety	NS	2.77	2.03
Aggressiveness	NS	1.26	0.85

(From Lazzari et al., 1975) (15).

presented a reduction of SAMe content in the brain, which could be attributed to an increased turnover of transmethylation processes.

So we carried out an open study of 51 patients diagnosed by Kilholz's classification as follows:

— Endogenous depression: 45 cases;
— Reactive depression: 1 case;
— Involutional depression: 2 cases;
— Neurotic depression: 3 cases (9).

The experiment was designed as a multicentric single-blind study with evaluation by the Hamilton Rating Scale for Depression (HRSD). The participating centers were various psychiatric hospitals in Northern and Central Italy. The patients were hospitalized with a diagnosis of depression confirmed by the Brief Psychiatric Rating Scale (BPRS) of Overall and Gorham; all had total HRSD scores in excess of 40. The HRSD was readministered at the end of the first and second week of treatment. SAMe was administered by personnel other than the attending physicians. The product was given in three intramuscular injections a day for a total of 45 mg daily. Patients with abnormal blood chemistry findings or with cardiac, renal, or hepatic pathology were not included in this study.

Table 2 shows the results, grouped in two separate lists according as the respective items showed marked improvement (more than 50%) or variations of 25% or less. A third heading was provided in the table for symptoms showing aggravation during the treatment; but as it turned out, none of the items showed aggravation and the heading remained blank.

It is interesting to note that the items that remained unchanged or were modified only to the extent of 25% or less were either not characteristic of depression or of minor import — incidentally, the marked amelioration of such symptoms may simply reflect the heterogeneous composition of the test group. Conversely, all the items that showed highly significant changes ($P < 0.001$ in the t-test for paired data) were typical of depression.

The latency of SAMe effects may be estimated by comparing the magnitude of improvement at 1 and 2 weeks of treatment (Table 3). Antidepressant effects from SAMe were already significant at the end of the first week; but the percentages of improvement were definitely higher at the end of the second week. In Table 3 we have underscored the values for items that are more characteristic of the depressive state, the better to show how these were modified more than the others. At any rate, the effect of SAMe on the "depressive nucleus" of the HRSD, as depicted in Table 4, shows a high incidence of improved patients along with good percentages of amelioration of separate symptoms.

If we regroup our results by an arbitrary grading of total HRSD scores into "excellent" (75% or more), "good" (not less than 50%), "fair" (not less than 25%), and "poor" (less than 25%), we find that the results were excellent in 37.5% of the cases, and good in 21.5%, for a total of 59% of the sample (patients showing clinical amelioration). The same percentages emerged from a strictly clinical assessment; and likewise, the incidence of patients who failed to show clinical evidence of improvement was the same as that graded as "poor" in Table 4.

So then, in short, SAMe exhibited prompt and marked antidepressant effects in these trials, with complete freedom from adverse side actions.

Table 2. HRSD items influenced by SAMe at 2 weeks of treatment in the open trial

HRSD items	% Patients improved	% Symptom improvement	Mean difference	t-Test
Symptoms improved (50% or more)				
Depressed mood	82	57	1.55	HS
Psychic anxiety	78	61	1.63	HS
Work and hobbies	72	55	1.33	HS
Insomnia	66	57	1.12	HS
Suicidal thoughts	62	43	0.79	HS
Somatic anxiety	62	49	0.93	HS
Hypochondria	54	49	1.23	HS
Symptoms not influenced (25% or less)				
Daytime symptom oscillations	27.5	63	0.85	HS
Guilt feelings	26	63	0.95	HS
Genital symptoms	26	46	0.73	HS
Obsessive symptoms	25	70	1.12	HS
Paranoid symptoms	16	59	1.18	NS
Depersonalization	8	55	0.83	NS
Disease awareness	2	7	0.09	NS
Symptoms aggravated		None .		

HS = P < 0.001   NS = P > 0.05

(From Agnoli et al., 1975) (2).

In regard to the latter, considering that classical antidepressants may produce certain side effects, we applied to our clinical material a Side Effects Rating Scale (SERS) of our own making; but we found no side effects at all, with the possible exception of a mild manic rebound in three patients, between the third and the fifth day of treatment. All these findings are in good agreement with published data (8, 11, 20).

After this open trial we started a double-blind trial of SAMe versus placebo, involving 30 randomly selected depressed patients. Twenty patients received SAMe in three intramuscular injections a day for a total of 45 mg daily, and the remaining 10 received a dummy medication indistinguishable from the active drug and administered with the same ceremonial; the duration of treatment varied from 5 to 30 days for a mean duration of 22 days. The criterion for determining the length of administration in each case was clinical improvement in terms of total HRSD ratings: treatment was discontinued when improvement reached 75% or more, and conversely no treatment was continued beyond 30 days (2, 17).

Table 3. Comparison of percent improvement of HRSD items at 1 and 2 weeks of treatment

Items	First week %	Significance	Second week %	Significance
1 Depressed mood	<u>37</u>	HS	<u>57</u>	HS
2 Guilt feelings	33	NS	63	HS
3 Suicidal thoughts	<u>57</u>	HS	<u>43</u>	HS
4 Insomnia	33	HS	57	HS
5 Work and hobbies	28	HS	<u>55</u>	HS
6 Retardation	<u>39</u>	HS	<u>60</u>	HS
7 Psychomotor agitation	45	HS	59	HS
8 Psychic anxiety	38	HS	61	HS
9 Somatic anxiety	35	HS	49	HS
10 Gastrointestinal somatic symptoms	35	HS	48	HS
11 General somatic symptoms	34	HS	47	HS
12 Genital symptoms	26	NS	46	HS
13 Hypochondria	33	HS	59	HS
14 Disease awareness	6	NS	7	NS
15 Depersonalization and derealization	55	NS	55	NS
16 Paranoid symptoms	54	NS	59	NS
17 Obsessive symptoms	55	S	70	HS
18 Daytime oscillations of symptoms	39	HS	63	HS

NS = P > 0.05   S = P < 0.05   HS = P < 0.01

(From Agnoli et al., 1975) (2).

Table 4. Effects of SAMe on the "depressive nucleus" of the HRSD at the end of 2 weeks of treatment

Items characteristic of depression	No. and % improved patients	% improvement of item
Depressed mood	42/51 (82.5%)	57
Suicidal thoughts	32/51 (62.7%)	43
Work initiative	37/51 (72.5%)	55
Retardation	32/51 (62.7%)	60

(From Agnoli et al., 1975) (2).

Figure 2 shows the percentages of patients improving with active medication and with placebo, again in terms of HRSD scores. Ostensibly, improvement involving the typical items of the "depressive nucleus" was more satisfactory in patients receiving SAMe than in those treated with placebo. Also, the results show that SAMe had a significant action on all the principal items, whereas the effects of placebo never reached significance.

Table 5 shows that the items more markedly modified by SAMe treatment were those more characteristic of depression, in terms of which more patients showed clinical amelioration.

Now in regard to the various forms of depression included in this study, both the overall improvement and the amelioration of the distinctive "depressive nucleus" were evenly distributed in the two main groups of patients (endogenous and reactive depression), even though the improvement expressed by total scores was greater in the endogenous depression group (Table 6). The interval between the start of treatment and the onset of clinical improvement was from 4 to 6 days.

If we regroup our patients by the same criterion as in our previous trial, we find that the results were excellent in 50% of the cases and good in 15% — roughly the same figures as in the open study (Table 7).

Still further confirmation of the antidepressant effects of SAMe came from another double-blind study versus imipramine (16), which was done seeking a better definition of the antidepressant spectrum of SAMe with reference to an established drug of known activity, such as imipramine.

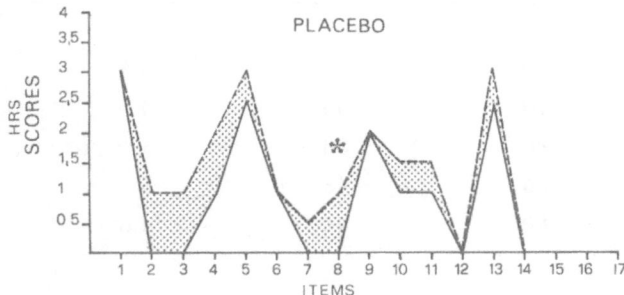

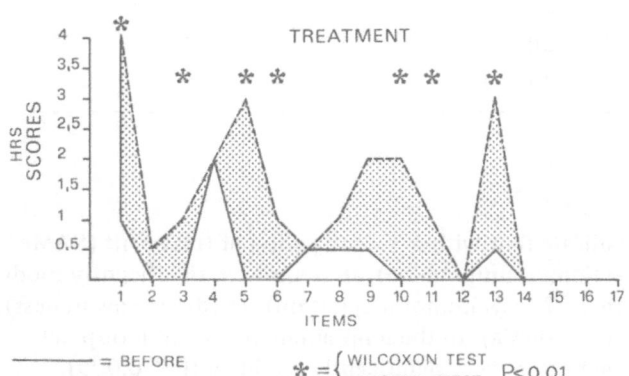

Fig. 2. Comparison of SAMe and placebo effects, before and after treatment: improvement index of symptoms for each item of the HRSD (mean values of HRSD scores). (From Agnoli et al., 1976) (1)

Table 5. Percent patients improved and percent item improvement of HRSD items more favorably influenced by SAMe in the double-blind study

ITEMS	Percent patients improved	Percent item improvement
Depressed mood	100	72.8
Work initiative	90	81
Somatic symptoms	85	80
Retardation	80	82
Suicidal thoughts	70	86
Hypochondria	65	66

(From Agnoli et al., 1975) (2).

Table 6. Comparison of SAMe and placebo in terms of clinical amelioration of various types of depressive syndromes, as expressed by the HRSD

Diagnosis	No. of patients	% improvement of depressive nucleus	% improvement of overall score	Mean latency of effect (days)
SAMe				
Endogenous depression	7	69	62	6.5
Reactive depression	6	70	46	4
Endoreactive depression	1	86	88	5
Involutional depression	1	66	48	5
Neurotic depression	5	80	44	4.6
Placebo				
Endogenous depression	3	25		
Neurotic depression	4	33		
Reactive depression	1	50		
Involutional depression	2	14		

(From Agnoli et al., 1976) (1).

The results shown in Table 8 indicate that with a 3-week period of treatment (SAMe 75 mg daily by intramuscular injection, or imipramine), all items were significantly modified by SAMe ($P < 0.05$) except items 7 (psychomotor agitation), 14 (disease awareness), and 18 (oscillations of mood and cenesthesia). In the imipramine treatment group, all the more characteristic items of depression were significantly modified ($P < 0.005$),

Table 7. Overall improvement as expressed by the HRSD total score in patients with different clinical forms of depression (double-blind study)

Improvement as assessed by HRSD scores	No. of patients	Clinical Forms				
		Endogenous depression	Reactive depression	Endoreactive depression	Involutional depression	Neurotic depression
75% (Excellent)	10 (50%)	4	2	1	0	3
50% (Good)	3 (15%)	1	2	0	0	0
25% (Fair)	6 (30%)	2	1	0	1	2
Less than 25% (Poor)	1 (5%)	0	0	0	0	1
Totals	20	7	5	1	1	6

(From Fazio et al., 1973) (9).

whereas items 7 (psychomotor agitation), 11 (general somatic symptoms), 12 (genital symptoms), 13 (hypochondria), 14 (disease awareness), and 18 (oscillations of mood and cenesthesia) were not significantly modified (Table 8).

In terms of direct action upon the depressive nucleus, both the percentages of score improvement and the number of patients improved were practically the same with SAMe and with imipramine (Table 9). In fact, with due allowance for the diversification of depression types, there seems to be no way to tell the effects of SAMe from those of imipramine, at least with the dosages employed and with the duration of treatment selected for these trials.

Table 8. Comparison of results obtained with SAMe and imipramine in depressed patients, as expressed by HRSD item scores (statistical analysis)

| Items | Wilcoxon test | |
	SAMe	Imipramine
1	S	S
2	S	S
3	S	S
4	S	S
5	S	S
6	S	S
7	NS	NS
8	S	S
9	S	S
10	S	S
11	S	NS
12	S	NS
13	S	NS
14	NS	NS
15	S	S
16	S	S
17	S	S
18	NS	NS
$S = P < 0.05$	$NS = P > 0.05$	

(From Mantero et al., 1975 (16).

Table 9. Effects of SAMe and imipramine on the "depressive nucleus". Percent patients improved and percent item improvement as expressed by the scores of HRSD item more favorably influenced by SAMe

Items	Percent patients improved	Percent item improvement
SAMe		
Depressed mood	100	72
Suicidal thoughts	92	87
Work initiative	87	63
Ratardation	81	73
Imipramine		
Depressed mood	100	68
Suicidal thoughts	86	70
Work initiative	93	59
Retardation	61	55

(From Mantero et al., 1975) (16).

Conclusions

From the results of our clinical trials with SAMe we may conclude that this substance performs clinically as an antidepressant drug. Its salient features are a complete absence of toxicity and freedom from side effects; a prompt antidepressant action (usually in 4 days of treatment in average cases), and a type of action directed quite selectively toward the symptoms that are most typical of depressive syndromes. The action is ostensibly that of a pure (nonanxiolytic) antidepressant. Good or excellent results were obtained in 59% of the cases in the single-blind trial and in 65% in the double-blind trial.

Separate symptoms showed different percentages of improvement. Improvement of items expressing the depressive nucleus (depressed mood, suicidal thoughts, loss of work initiative, retardation) was more significant in the double-blind trial (80% on average) than in the open study (54%).

Preliminary tests in normal volunteers indicated no psychostimulant activity from SAMe; the antiasthenic effects of the substance should be explored in greater depth.

We should like to amplify the study of previous findings concerning the deterioration of symptoms in patients with acute schizophrenia, which reproposes the issue of trans-methylation processes in the CNS and of the methylation of putative mediators of behavior. Therapeutically, we feel that SAMe should be tried clinically in torpid forms of schizophrenia in order to test its "unblocking" or activating effect seen in acute forms of the psychosis.

References

1. Agnoli, A., Andreoli, V., Casacchia, M., Cerbo, R.: Effect of S-adenosyl-L-methionine (SAMe) upon depressive symptoms. J. psychiat. Res. **13** (1), 43 (1976)
2. Agnoli, A., Fazio, C., Andreoli, V., De Carolis, U., Bonamini, F., Pastorino, P., Casacchia, M., Cerbo, R., Ruggeri, S., Carolei, A.: Disturbi neuropsichiatrici e transmethilazioni. Effetti terapeutici della S-adenosil-L-metionina (SAMe). Clin. Ter. **75**(6), 567 (1975)
3. Antun, F.T.: The effects of L-methionine (without IMAO) in schizophrenia. J. psychiat. Res. **8**, 63 (1971)
4. Axelrod, J., Cohn, C.K.: Methyltransferase enzymes in red blood cells. J. Pharmacol. exp. Ther. **176**, 650 (1971)
5. Baldessarini, R.J.: Alterations in tissue levels of S-adenosylmethionine. Biochem. Pharmacol. **15**, 741 (1966)
6. Boquist, L.: The effect of excess methionine on the pancreas. Lab. Invest. **21**, 96 (1969)
7. Cohen, H.P., Berg, C.P.: Response of rats to diets high in methionine. Fed. Proc. **10**, 172 (1951)
8. Editorial: A possible mechanism for methionine toxicity in guinea pigs. Nutr. Rev. **28**, 293 (1970)
9. Fazio, C., Andreoli, V., Agnoli, A., Casacchia, M., Cerbo, R.: Effetti terapeutici e meccanismo d'azione della S-adenosil-L-metionina (SAMe) nelle sindromi depressive. Minerva med. (Torino) **64**, 1515 (1973)
10. Fazio, C., Andreoli, V., Agnoli, A., Casacchia, M., Carbo, R., Pinzello, A.: Therapy of schizophrenia and depressive disorders with S-adenosyl-L-methionine. Intern. Res. Comm. System (IRCS), Clin. Pharmacol. Ther. **2**, 1015 (1974)
11. Hardwick, D.F., Applegarth, D.A., Cockcroft, D.M., Ross, P.M., Calder, R.J.: Pathogenesis of methionine induced toxicity. Metabolism **19**, 381 (1970)
12. Jeanjean, M., Taper, H.: Toxicité de la methionine chez le lapin. Path. europ. **2**, 93 (1967)
13. Klavins, J.V., Kinney, T.D., Kaufman, N.: Histopathologic changes in methionine excess. Arch. Path. (Chicago) **75**, 99 (1963)
14. Labò, G., Gasbarrini, G., Miglio, F.: Alcuni effetti delle transmetilazioni SAMe-dipendenti in epatologia. Minerva med. (Torino) **63** (35), 2007 (1972)
15. Lazzari, R., Agnoli, A., Cerbo, R., Borgo, S. Casacchia, M., Ruggeri, S.: The effect of so-called non-analeptics on the mood of normal subjects: the effect of SAMe. Psychopharmacologia. In press (1977)
16. Mantero, M., Pastorino, P., Carolei, A., Agnoli, A.: Studio controllato in doppio cieco (SAMe-Imipramina) nelle sindromi depressive. Minerva med. (Torino) **66**, 4098 (1975)
17. Nowlys, U.: Methods for studying mood changes produced by drugs. Rev. Psychol. appl. **11**, 373 (1961)
18. Pollin, W., Cardon, P.V., Kety, S.S.: Effects of amino acid feedings in schizophrenic patients treated with iproniazid. Science **133**, 104 (1961)
19. Stramentinoli, G., Pezzoli, C., Catto, E.: Aspetti farmacologici della S-adenosil-L-methionina in epatologia. a) Cenni biochimici; b) Tossicità acuta, cronica e fetale; c) Modificazioni biochimiche e morfoistochimiche in ratti controllo e trattati con S-adenosil-L-methionina dopo intossicazione sperimentale da vari agenti. Minerva med. (Torino) **66**, 1541 (1975)
20. Szantay, I., Szirmai, E., Cotul, S.: Metabolism de la ^{35}S-methionine dans la cirrhose hepatique. Presse med. **79**, 399 (1971)
21. Williams, J.N., Jasik, A.D.: A markedly antilipotropic action of methionine. Nature (Lond.) **200**, 472 (1973)

Subject Index

Monographien aus dem Gesamtgebiete der Psychiatrie
Psychiatry Series

Herausgeber: H. Hippius, W. Janzarik, M. Müller

1. Band: K. Hartmann
Theoretische und empirische Beiträge zur Verwahrlosungsforschung
2., neubearbeitete und erweiterte Auflage. 1977. 16 Abbildungen, 34 Tabellen. XII, 180 Seiten
ISBN 3-540-07925-4

2. Band: P. Matussek
Die Konzentrationslagerhaft und ihre Folgen
Mit R. Grigat, H. Haiböck, G. Halbach, R. Kemmler, D. Mantell, A. Triebel, M. Vardy, G. Wedel
1971. 19 Abbildungen, 73 Tabellen. X, 272 Seiten
ISBN 3-540-05214-3

3. Band: A. E. Adams
Informationstheorie und Psychopathologie des Gedächtnisses
Methodische Beiträge zur experimentellen und klinischen Beurteilung mnestischer Leistungen
1971. 12 Abbildungen.
IX, 124 Seiten
ISBN 3-540-05215-1

4. Band: G. Nissen
Depressive Syndrome im Kindes- und Jugendalter
Beitrag zur Symptomatologie, Genese und Prognose
1971. 11 Abbildungen, 51 Tabellen. IX, 174 Seiten
ISBN 3-540-05493-6

5. Band: A. Moser
Die langfristige Entwicklung Oligophrener
Mit einem Vorwort von Chr. Müller
1971. 4 Abbildungen, 30 Tabellen. X, 102 Seiten
ISBN 3-540-05599-1

6. Band: H. Feldmann
Hypochondrie
Leibbezogenheit. Risikoverhalten. Entwicklungsdynamik
1972. 36 Abbildungen, 5 Tabellen. VI, 118 Seiten
ISBN 3-540-05753-6

7. Band: S. Meyer-Osterkamp, R. Cohen
Zur Größenkonstanz bei Schizophrenen
Eine experimentalpsychologische Untersuchung. Mit einem einführenden Geleitwort von H. Heimann
1973. 5 Abbildungen.
VII, 91 Seiten
ISBN 3-540-06147-9

8. Band: K. Diebold
Die erblichen myoklonisch-epileptisch-dementiellen Kernsyndrome
Progressive Myoklonusepilepsien – Dyssinergia cerebellaris myoclonica – myoklonische Varianten der drei nachinfantilen Formen der amaurotischen Idiotie
1973. 31 Abbildungen.
IX, 254 Seiten
ISBN 3-540-06117-7

9. Band: C. Eggers
Verlaufsweisen kindlicher und präpuberaler Schizophrenien
1973. 3 Abbildungen.
IX, 250 Seiten
ISBN 3-540-06163-0

10. Band: M. Schrenk
Über den Umgag mit Geisteskranken
Die Entwicklung der psychiatrischen Therapie vom „moralischen Regime" in England und Frankreich zu den „psychischen Curmethoden" in Deutschland
1973. 20 Abbildungen.
IX, 194 Seiten
ISBN 3-540-06267-X

11. Band: Heinz Schepank
Erb- und Umweltfaktoren bei Neurosen
Tiefenpsychologische Untersuchungen an 50 Zwillingspaaren
Unter Mitarbeit von P. E. Becker, A. Heigl-Evers, C. O. Köhler, Helga Schepank, G. Wagner
1974. 1 Abbildung, 82 Tabellen.
VIII, 227 Seiten
ISBN 3-540-06647-0

12. Band: L. Ciompi, C. Müller
Lebensweg und Alter der Schizophrenen
Eine katamnestische Langzeitstudie bis ins Senium
27 Fallbeispiele.

1976. 23 Abbildungen, 48 Tabellen. IX, 242 Seiten
ISBN 3-540-07567-4

13. Band: L. Süllwold
Symptome schizophrener Erkrankungen
Uncharakteristische Basisstörungen
1977. 15 Tabellen.
VIII, 112 Seiten
ISBN 3-540-08203-4

14. Band: **The Appalic Syndrome**
Editors: G. Dalle Ore, F. Gerstenbrand, C. H. Lücking, G. Peters, U. H. Peters
With the editorial assistance of E. Rothemund
1977. 67 figures, 17 tables.
XV, 259 pages
ISBN 3-540-08301-4

15. Band: O. Benkert
Sexuelle Impotenz
Neuroendokrinologische und pharmakotherapeutische Untersuchungen
1977. 33 Abbildungen, 20 Tabellen. VIII, 139 Seiten
ISBN 3-540-08427-4

16. Band: R. Avenarius
Der Größenwahn
Erscheinungsbilder und Entstehungsweise
1978. VI, 98 Seiten
ISBN 3-540-08547-5

17. Band: **Psychiatrische Epidemiologie**
Geschichte, Einführung und ausgewählte Forschungsergebnisse
Herausgeber: H. Häfner
1978. 20 Abbildungen, 91 Tabellen. X, 252 Seiten
ISBN 3-540-08629-3

18. Band: **Transmethylations and the Central Nervous System**
Edited by V. M. Andreoli, A. Agnoli, C. Fazio
1978. 45 figures, 42 tables.
Approx. 220 pages
ISBN 3-540-08693-5

19. Band: **Psychiatrische Therapie-Forschung**
Ethische und juristische Probleme
Herausgeber: H. Helmchen, B. Müller-Oerlinghausen
1978. Etwa 165 Seiten
ISBN 3-540-08732-X

In Vorbereitung
R. M. Torack
The Pathological-Physiology of Demetia

Preisänderungen vorbehalten

Karl R. Popper
Penn, Great Britain

John C. Eccles
Contra, Switzerland

The Self and Its Brain

1977. 66 figures. XVI, 597 pages.
ISBN 3-540-08307-3

Contents:
Materialism Transcends Itself. The Worlds 1,2 and 3. Materialism Criticized. Some Remarks on the Self. Historical Comments on the Mind-Body Problem. Summary. – The Cerebral Cortex. Conscoius Perception. Voluntary Movement. The Language Centres of the Human Brain. Global Lesions of the Human Cerebrum. Circumscribed Cerebral Lesions. – The Self-Conscoius Mind and the Brain. Conscious Memory: The Cerebral Processes Concerned in Storage and Retrieval. – Dialogues between the two authors.

This book iss timely, as it appears at a point of impasse between philosophy and science. It creates the first link between the philosophy of the self and neurobiology. In dealing with the self, philosophers have so far taken little account of scientific knowledge of the brain; scientists, for their part, have traditionally avoided philosophy in favour of purely material evidence.

Eccles (a neurobiologist) and Popper (a philosopher), both believers in dualism and interactionism, consider the existence of consciousness one of the greatest riddles of cosmology.

Part I, Popper discusses the philosophical issue between dualist or even pluralist interactionism on the one side, and materialism and parallelism on the other. There is also a historical review of these issues.

In Part II, Eccles examines the mind from the neurological standpoint: the structure of the brain and its functional performance under normal as well as abnormal circumstances, for example when lesions (especially those surgically induced) are present. The result is a radical and intriguing hypothesis on the interaction between mental events and detailed neurological occurrences in the cerebral cortex.

Part III, based on twelve recorded conversations, reflects the exciting exchange between the authors as they attempt to come to terms with their conflicting opinions. This part preserves the intimate quality of these dialogues, ans shows how some of the authors' viewpoints changed in the course of these daily discussions.

Springer
International